Challenges in Physics Education

This book series covers the many facets of physics teaching and learning at all educational levels and in all learning environments. The respective volumes address a wide range of topics, including (but not limited to) innovative approaches and pedagogical strategies for physics education; the development of effective methods to integrate multimedia into physics education or teaching/learning; innovative lab experiments; and the use of web-based interactive activities. Both research and experienced practice will feature prominently throughout.

The series is published in cooperation with GIREP, the International Research Group on Physics Teaching, and will include selected papers from internationally renowned experts, as well as monographs. Book proposals from other sources are entirely welcome.

Challenges in Physics Education addresses professionals, teachers, researchers, instructors and curriculum developers alike, with the aim of improving physics teaching and learning, and thereby the overall standing of physics in society.

Book proposals for this series may be submitted to the Publishing Editor: Marina Forlizzi; email: Marina.Forlizzi@springer.com

More information about this series at http://www.springer.com/series/16575

Beata Jarosievitz · Csaba Sükösd
Editors

Teaching-Learning Contemporary Physics

From Research to Practice

Springer

Editors
Beata Jarosievitz
Dennis Gabor College
Budapest, Hungary

Csaba Sükösd
Budapest University of Technology
and Economics
Budapest, Hungary

ISSN 2662-8422 ISSN 2662-8430 (electronic)
Challenges in Physics Education
ISBN 978-3-030-78719-6 ISBN 978-3-030-78720-2 (eBook)
https://doi.org/10.1007/978-3-030-78720-2

This Springer imprint is published by the registered company Springer Nature Switzerland AG
The registered company address is: Gewerbestrasse 11, 6330 Cham, Switzerland

Preface

It is generally accepted that the greatest achievement in physics since Newton was the discovery of the quantum behavior of the microscopic world and the discovery of the structure and behavior of the space–time. The first one led to the development of the quantum physics and the second to the theory of relativity. These groundbreaking discoveries have completely changed not only our views about Nature, but also our everyday life.

However, despite of the utmost importance of these physics theories, their introduction into the public education is lagging behind. The Physics Education Research (PER) has already carried out researches on teaching/learning quantum physics (QP) from many perspectives, focusing on specific aspects and concepts or developing didactic approaches and educational paths. There is still a great debate about which contents have to be proposed, which didactic approaches and strategies have to be adopted, and which level of formalization is possible to include. There are probably more proposals on the teaching of QP at high-school level than those for classical mechanics. Despite of this, there is still a lack of resources available for teachers, for school practice and for a coherent integration of QP in the curriculum.

GIREP vzw—the International Research Group on Physics Teaching—has established a *Community on Teaching/Learning Quantum Physics* to improve the situation above and asked internationally renowned researchers to join. This community had its very first meeting and issued a statement during the GIREP–ICPE–EPEC–MPTL 2019 Conference held in Budapest (Hungary) on July 1–5, 2019. The main focus of this book is to present research papers about the introduction of contemporary physics topics—mainly but not exclusively quantum physics—into school. The papers are selected from the contributions of the conference mentioned above. All of the 19 papers of the book show results of genuine new research works dealing with Physics Education.

The papers of the book are grouped into five parts. Below, we give a short summary of their content.

Part One—Teaching/Learning Quantum Physics

This part comprises five chapters, which give a broad overview on the different aspects and approaches of teaching quantum phenomena at different educational levels. An overview of the literature in the field is also presented, framing the contributions in three main approaches: historical approach; formal structural approach; and conceptual approach. The main features of each setup are discussed on the basis of an extensive research bibliography.

Quantum physics is not only a fundamental physical theory but recently promises big technological advances using the so-called quantum information. These developments should have its influence on teaching physics at school level in order to give students insight into a fascinating and fundamental part of modern physics. Herewith, they can experience the fundamental notions of quantum physics: non-determinism, superposition, and uncertainty.

Mathematics and sciences, in particular physics, are structurally related to each other. Therefore, when talking about quantum physics, it is also necessary to deal with its relationship to mathematics. A research is presented which analyzes the spontaneous ideas about the relationship between math and physics of prospective primary teachers. There is also a paper discussing the modes and means for introduction of contemporary physics to pre-university-level education illustrated by a few successful examples. Examples come from soft matter physics in liquid crystals and hydrogels and solid-state physics in superconductors.

Part Two—Roland Eötvös and the Equivalence Principle

The year 2019 was an UNESCO-endorsed commemorative year on the achievements of Roland Eötvös Hungarian physicist. With his extremely sensitive torsion balance, he verified experimentally the strict proportionality between the inert mass and gravitational mass with an unprecedented accuracy, which was one of the basic starting points of the general relativity of Albert Einstein. This part contains two papers dealing with Eötvös and the equivalence principle.

A Hungarian professor recalls the memory of Baron Roland Eötvös, an outstanding figure of the experimental exploration of the gravitational interaction and "funding father" of applied geophysics. Another essay briefly depicts the history of the concept of weight–gravity and its impact on physics teaching in regular school classes. What makes history especially interesting is that the debate on weight definition in physics teaching is not closed and, in fact, splits the community of physics educators around the globe into two groups: Newtonians and Einsteinians.

Part Three—Experimentation, Impact of PER and Assessment

Experiments are essential in physics research; therefore, experiments should also be essential in the physics teaching/learning process. The learning experiences are given different names, experiments, practicals, investigations and, in the current context, can be referred to as inquiry-based learning and project-based learning to STEM projects. In whatever form, curricula and pedagogies internationally are aspiring to instill the wonder of science through such pedagogies.

An important aspect of physics education research is understanding how students use their prior knowledge in making sense of the concepts of physics that they are studying. In these situations, the students must transfer knowledge which was acquired in a different context to a situation in physics. However, transfer *within* physics learning is equally important. Students need to be able to use knowledge from one part to learn successfully other concepts.

An important part of experiments is the ones on thermal phenomena demonstrating the irreversibility. In the approach described, the term "laboratory" is not intended as referring solely the physics laboratory, but as an integrated laboratory perspective, comprising physical experiments, game with toy models, computer simulation, group discussion, and group tasks.

Lawson's Classroom Test of Scientific Reasoning (LCTSR) is a widely used two-tier test in physics education. Results show that statistically indistinguishable person measures were obtained, but that the choice of scoring method impacted test length and targeting and, therefore, also reliability and standard errors of person measures

Another study describes the design and implementation of a professional development program for high-school teachers, aimed at improving their competences in the use of practical work for the teaching and learning of physics. During the program, teachers also carried on their own action research projects.

Force Concept Inventory is a multiple-choice questionnaire commonly used to assess students' conceptual understanding of Newtonian mechanics. Such an analysis of student answers gave insights into the relationships between the student ideas about the force concept and their ability to correctly answer questions involving the first and second Newton's law.

Part Four—Active Learning

The Activity-Based Physics Group is best known for a number of physics active learning curricular developments including *Real Time Physics*, *Interactive Lecture Demonstrations*, and *Workshop Physics*. A paper describes the context of these programs and the characteristics that have made them successful and sustainable over the years.

Over the last several years, active learning methods and strategies have received considerable attention from the educational community and are commonly presented in the related literature as a credible solution to the reported lack of efficacy of more "traditional" educative approaches. Research has shown that a possible factor is the strongly contextualized nature of active learning that focuses on the interdependence of situation and cognition. A paper presents the results of a symposium with different contributions in the field of research on active learning.

Part Five—Innovative Projects

Innovation is the driving force behind the advancement of almost every field of society. Similarly, innovative projects are important also for the teaching/learning process.

One of such innovative example is Max's Worlds, which is a laboratory project aimed at developing a coherent vertical curriculum suitable for kindergarten and primary school teachers, either for their pre-service university program or for in-service training activities. The project provides didactic suitcases.

The development of creativity in students is a critical objective of renewed science curricula. A study describes an after-school program, called *TECNOArtea*, designed for the development of double exceptionality students' creativity through problem-based science sessions.

In secondary education, cathode-ray tubes (CRTs) are often the first choice when it comes to investigating the behavior of electrically charged particles in electric and magnetic fields. While CRTs offer some advantages, mainly from a practical point of view, they are on the whole ill-suited for an inquiry-based approach since they provide very limited room for modification or hands-on experimentation. Therefore, an innovative 3D printable plasma electron gun has been developed, which is at the same time modular, inexpensive, and easily accessible.

The last paper proposes a novel teaching/learning sequence on tribology, based on the experimental investigation of Gecko®, a bio-inspired microstructured synthetic material with peculiar tribological properties. The sequence has been tested on a group of honors high-school students with encouraging results.

Budapest, Hungary

Beata Jarosievitz
Csaba Sükösd

Acknowledgements The editors would like to thank all authors of this book for their cooperation in preparing the manuscript. Also, they are indebted to GIREP—especially Prof. Marisa Michelini GIREP president—for the continuous encouragement and support. Finally, they express their thanks for the staff of Springer for their valuable advices and help in the preparation of this book.

Contents

Teaching/Learning Quantum Physics

Approaches on T/L Quantum Physics from PER Literature 3
Marisa Michelini and Alberto Stefanel

Quantum Cryptography as an Approach for Teaching Quantum Physics 19
Gesche Pospiech

Milq—Quantum Physics in Secondary School 33
Rainer Müller and Oxana Mishina

The Role of Mathematics in Teaching Quantum Physics at High School 47
Gesche Pospiech, Avraham Merzel, Giacomo Zuccarini, Efi Weissman, Nadav Katz, Igal Galili, Lorenzo Santi, and Marisa Michelini

Introduction of Contemporary Physics to Pre-university Education 71
Jerneja Pavlin, Alberto Stefanel, Philippe Lindenau, Michael Kobel, Anja Kranjc Horvat, Jeff Wiener, Sascha Schmeling, Andreas Borowski, Dagmara Sokołowska, and Mojca Čepič

Roland Eötvös and the Equivalence Principle

Roland Eötvös: Scientist, Statesman and Educator 93
András Patkós

The Impact of the Equivalence Principle on Physics Teaching—The Ongoing Opposition in Teaching of Weight-Gravity 109
Igal Galili

Experimentation, Impact of PER and Assessment

Experimentation in Physics Education: Should We Bother 127
Manjula D. Sharma

Context and Transfer: How Physics Education Research Informs Teaching and Learning 137
Dean Zollman

Teaching Thermal Phenomena and Irreversibility Through Playable Dice and Coin Toy Models 149
Massimiliano Malgieri, Anna De Ambrosis, Monica Di Savino, and Pasquale Onorato

Measuring Scientific Reasoning Using the LCTSR 163
Maja Planinic, Ana Susac, Lana Ivanjek, Katarina Jelicic, and Karolina Matejak Cvenic

A Community of Learners on Laboratory Work. Design and Implementation of a Teacher Training Programme 171
Marta Carli and Ornella Pantano

A Study on Engineering Freshman Conceptual Understanding of Newtonian Mechanics 185
Onofrio Rosario Battaglia and Claudio Fazio

Active Learning

Active Dissemination—Over Three Decades of Faculty Development in Active Learning 201
David R. Sokoloff

Strategies for Active Learning to Improve Student Learning and Attitudes Towards Physics 213
Claudio Fazio, Marina Carpineti, Sergej Faletič, Marco Giliberti, Gareth Jones, Eilish Mcloughlin, Gorazd Planinšič, and Onofrio Rosario Battaglia

Innovative Projects

Max's Worlds: An Innovative Project for K-6 Science Education 237
Federico Corni and Liliana Dozza

Creativity Development Through Problem-Based Informal Science: The Case of Double Exceptionality Students 251
Martha Lucia Orozco-Gomez, Radu Bogdan Toma, and Maria Merino

3D-Printed Plasma Cathode Electron Source for Educational Purposes 263
Fabian Bernstein, Sascha Schmeling, Thomas Wilhelm, and Julia Woithe

The Gecko® Approach to Friction: A Novel Teaching Learning Sequence 273
Cinzia Scorzoni, Guido Goldoni, and Valentina De Renzi

Teaching/Learning Quantum Physics

Approaches on T/L Quantum Physics from PER Literature

Marisa Michelini and Alberto Stefanel

Abstract Quantum mechanics has a paradigmatic role in twentieth-century physics for the theoretical foundation of the description of the world and important application and technological implications relevant also in everyday life. The Physics Education Research (PER) from many decades carried out researches on Teaching/Learning Quantum Physics (QP) from many perspectives, focusing on specific aspects and concepts or developing didactic approaches and educational paths. There is a great debate about which contents have to be proposed, which didactic approaches and strategies have to be adopted, which level of formalization is possible to include. The proposals on the teaching of QP at the high school level are perhaps more than those for classical mechanics. However, there is a lack of resources available for teachers, for school practice and for a coherent integration of QP in the curriculum. Here we propose an overview of literature in the field framing the contributions in three main approaches based on the perspective adopted for the approaches in organizing contents: historical approach; formal structural approach; conceptual. The main features of each set-up are discussed on the basis of an extensive research bibliography.

Keywords Educational approaches to quantum physics · High school · Physics education research literature

1 Introduction

Quantum Mechanics (QM) is the cultural paradigm of reference for the construction of the twentieth century for the current description of the world and in particular is the foundation of theoretical physics (Hadzidaki et al. 2000; Weissman et al. 2019; AAVV

M. Michelini (✉) · A. Stefanel
University of Udine, via delle Scienze 206, Udine, Italy
e-mail: marisa.michelini@uniud.it

A. Stefanel
e-mail: alberto.stefanel@uniud.it

B. Jarosievitz and C. Sükösd (eds.), *Teaching-Learning Contemporary Physics*, Challenges in Physics Education,
https://doi.org/10.1007/978-3-030-78720-2_1

2000, 2002). The new idea of state, of measurement and of the superposition principle changes the way of thinking and interpreting the microscopic world (Sakurai 1990; Mermin 1990, 1998; Pospiech 1999, 2000, 2003; Newton 2004; Ghirardi 2004). QM therefore occupies a special place inside the modern physics theories. Its implications are very important in many branches of science, not only for those in physics and chemistry, but also those in materials science, nanotechnologies, computation (Mermin 2003; Dür and Heuslery 2013); and in some advanced research lines in biophysics (Brookes 2017). There are many technological implications important also in everyday life (just think of laser or LED or in general to light sources and the emission processes as well as the way to improve their efficiency) (Zollman 2002; Johansson and Nilsson 2000). There is in particular a historical-epistemological value of quantum physics related to foundation of theoretical thinking (Hadzidaki et al. 2000; Weissman et al. 2019; Mermin 1998; Pospiech 2003; Ghirardi 2004). A relevant social and educational contribution is how QM offers the opportunity to develop formal thinking, by means of the discussion of the interpretative models of the microscopic world having only indirect information (Pospiech 1999, 2000, 2003; Michelini 2008; Francia 1975; Pospiech et al. 2008). The relevance of quantum interpretation of the world in physics and the relative applications, its cultural value and its educational opportunities motivate the PER for research on proposals to introduce it in secondary school curricula (AAVV 2000, 2002; Pospiech et al. 2008; Zollmann 1999).

There is a wide debate in literature on purpose and approach of the organization of contents (For the citizen's cultural education? For guidance? For dissemination?); what to deal with (the problems? the technologies? the applications?), how to deal with it (with a qualitative narrative …? Integrating it into classical physics or (at the opposite) emphasizing the elements of breakdown on the epistemological level?), to whom to propose it (to all students? Only to schools with a scientific curriculum?) (Pospiech et al. 2008; Krijtenburg-Lewerissa et al. 2019).

The different possible formulations and interpretations of the QM (Styer et al. 2002) have been envisaged as perspectives for different didactic proposals, that are more differentiated than those for classical mechanics (AAVV 2000, 2002; Pospiech et al. 2008; Cataloglu and Robinett 2002; Greca and Freire 2003; Stefanel 2008; Akarsu 2011; Lautesse et al. 2015; Baily and Finkelstein 2014). A wide literature documents the students' learning impact (Zollmann 1999; Cataloglu and Robinett 2002; Fischler and Lichtfeldt 1992; Müller and Wiesner 2002; Singh 2008; Wan et al. 2019). In particular, it was highlighted that the interpretation choices adopted in the quantum physics pathways affect the way in which students understand quantum concepts (Baily and Finkelstein 2014) and that students use different ontologies (Hoehn et al. 2019) and approach the concepts by different theoretical perspectives (Michelini and Stefanel 2008).

Several paths have been proposed: analysis of specific contexts, such as the interference (Fischler and Lichtfeldt 1992; Müller and Wiesner 2002) polarization of light or the spin of particles, for the construction of concepts/ideas (Pospiech 1999, 2000; Michelini 2008; French et al. 1975; Ghirardi et al. 1996; Michelini et al. 2000; McIntyre 2002, 2012; Kohnle et al. 2015); visualization as anchor for the intuition in the construction of quantum concepts, in two states systems (French et al. 1975; Ghirardi et al. 1996; Michelini et al. 2000; McIntyre 2002, 2012; Kohnle et al. 2015; Faletič and Kranic 2019) or more complex including energy levels, bands, tunneling (Zollman 2002; Cataloglu and Robinett 2002; Lawrence 1996; Redish et al. 2001; Herrmann 2000; Broklova and Koupil 2007); argumentation for interpretations with problem-based learning method (Müller and Wiesner 2002; Mason et al. 2015; Niedderer and Deylitz 1998; Petri and Niedderer 1998; Budde et al. 2002a, b); discussion of concepts and philosophical-epistemological analysis (Mermin 1990, 1998; Pospiech 1999, 2000, 2003); introduction of a reference formalism and discussion of its meaning (Francia 1975; Taylor 1998; Taylor et al. 1998; Ebison 1975; Haber-Schaim 1975). There is no consensus on the aspects to be treated and the approaches to be adopted (AAVV 2000,2002; Pospiech et al. 2008; Krijtenburg-Lewerissa et al. 2019; Lautesse et al. 2015), whether to confine oneself to non-relativistic quantum mechanics or to adopt approaches to field physics (Hobson 2005), whether to introduce the formalism and, eventually, which formalism to use (function of wave, Feynman paths, matrices, vector spaces.) (Francia 1975; Müller and Wiesner 2002; Wan et al. 2019; Michelini et al. 2000; Stadermann et al. 2019).

As far the curricula and experiences in schools are concerned, the choice is to introduce elements of quantum physics in almost all high school curricula in Europe and in many countries both in Asia and in America (Pospiech et al. 2008; Stefanel 2008; Akarsu 2011; Lautesse et al. 2015; Stadermann et al. 2019). For example, in the Italian guidelines from 2010 (decreto_indicazioni_nazionali 2010) is suggested to treat "The affirmation of the quantum light model [...] through the study of thermal radiation and the Planck hypothesis (addressed eventually only in a qualitative way)", the study of the photoelectric effect and the interpretation of Einstein, the theories and the experimental results evidencing discrete energy levels in the atom, the experimental evidence of the wave nature of matter, postulated by De Broglie, and the uncertainty principle "could significantly end the path". These topics appear in textbooks, even if often as an appendix of the classical physics or in a fragmented and non-organic way (Stadermann et al. 2019). Physics of quanta, quantum physics, quantum mechanics are often not identified and distinguished. Physics of quanta, although remembered as the old quantum "theory", does not have the coherence and completeness of a theory and, above all, it is not the foundation of quantum mechanics or of the more general quantum physics, which reproduce its results but in another theoretical and epistemological frame-works. Therefore, there is a need to produce awareness of the reference assumptions of the new mechanics, to offer indications on the formalism that is adopted and that assumes in the quantum mechanics a conceptual role. As Ghirardi points out, "there is a paradigm shift from classical to quantum physics. The change of perspective required by the new theory has conceptual implications that require epistemological reflection" (Ghirardi 2004). For that goal, teachers do not receive specific preparation on teaching learning QM. Moreover, resources for a coherent path of QM, integrated in physics curriculum are not available: the important contribution of PER in literature is rarely known by teachers and often it is specific to a cluster of concepts, students' difficulties and/or specific approaches.

There is a need for a profound research reflection that the PER must make starting from value, but also from the limits of the contributions it has proposed. As a contribution for this debate, here we summarize these research contributions, looking at how the contents are organized on the disciplinary level. Three main content approaches can be identified (Pospiech et al. 2008; Stefanel 2008): (1) Historical approaches, based on the discussion of interpretative problems analyzed in a historical key, more often proposed as a rational reconstruction of the historical development of ideas, through the transformation of crucial experiments and the reconstruction of the birth of the old quantum theory; (2) Formal structural axiomatic approach, based on the assumption of a formal description of the state of a quantum system to analyze the consequences on the interpretative level, generally proposed according to the wave formulation of quantum mechanics; (3) Conceptual approach, which introduces the founding concepts of quantum mechanics making them plausible in the analysis of specific phenomenological contexts (Pospiech et al. 2008; Stefanel 2008; Michelini et al. 2014).

These approaches have been adapted to different educational strategies and to different learning paths. The basic assumptions and the main limits are here discussed, as well as examples from literature.

2 Historical Approach

The analysis of the historical development of the interpretative problems that led to the crisis of classical physics and to the birth of quantum physics has an undoubted cultural value, that can provide an overview of the issues, which allows interesting and significant interdisciplinary links and developments (Hadzidaki et al. 2000; Weissman et al. 2019). This approach, perhaps, finds the best expression in the treatment of Born in his text Atomic Physics (Born 1989) and characterizes other classical texts (Messiah 1961). In the Born's proposal, the interpretative difficulties, which from time to time arise, are progressively overcome with new hypotheses, finding reconciliation with the previous conceptual framework at a higher level of understanding. Apparently contradictory concepts at the previous level, find a coherent placement at the next level. The innovative nature of QM emerges only for those aspects that can be somehow understood with descriptive categories of classical physics (for instance the relevance of discreteness of atomic level).

While ignoring these conceptual issues, an approach of this type, however, can be hardly translated into an educational path that can be easily proposed in high school. The main causes are the excessive time required, and above all the conceptual and formal pitfalls that should be addressed in any case. For these reasons it has been generally transposed to school (in many textbooks) as a narrative on the qualitative level of contents, acceptable, perhaps on a popularization level, but not on a didactic level (Messiah 1961). To try to overcome this limitation, a rational reconstruction of historical developments has been proposed through the analysis of crucial experiments (black body radiation, photoelectric effect, Compton effect, Zeeman effect…)

and the construction step by step of the old quantum theory, i.e., the quantization (discretization) of the different physical quantities used in the description of the systems (energy, angular momentum, spin …) (Klassen 2011; Niaz et al. 2010). The experimental approach constitutes at the same time its value and limit, because the analysis of the experimental aspects can prevail on the conceptual discussion of the theory that is going to build. Moreover, this approach is time spending, does not solve the problems related to the contradictory nature of physics of quanta, nor those of formalism difficulties. Teachers cannot avoid to adopt a descriptive and narrative treatment of the experiments set-up, results and explanations (Viau et al. 2011). The disciplinary contents are neglected, putting emphasis on the quantization hypotheses made and not discussing the problematic contexts in which these hypotheses have matured. The unifying dimension of theoretical thought is completely lost.

Most of the scholastic texts suggest reductions in these approaches. As has been highlighted by a recent study (Stadermann et al. 2019), the following aspects unite the texts for schools both in Europe, in Canada and Australia:

- physics of quanta is the preferred topics
- the historical approach is privileged or prevalent content
- the epistemological aspects considered important but not treated so much
- the technological applications are often included
- quantum physics is characterized by the following topics:
  - discrete atomic energy levels, interactions between light and matter
  - wave-particle duality
  - de Broglie wavelength
  - Heisenberg's uncertainty principle, and the probabilistic nature of quantum physics

It is important to take into account what is included in school textbooks to determine what is actually done in schools, considering that 50% of teachers offer their students nothing but what is written in the adopted textbooks, as it has emerged from the results of the TIMMS survey (Mullis et al. 2016).

3 Formal Structural Approach

Several researchers, recognizing the importance of formalism in QM and the need for a more direct approach to the theory than the historical approach allows, have developed proposals in which the mathematics on which the QM is based is constructed in specific classical contexts. These contexts are for instance oscillating systems, such as cords and membranes (Niedderer and Deylitz 1998; Petri and Niedderer 1998; Faletič 2015), or abstract systems of N coupled oscillators (Giannelli and Tarsitani 2005), or the interference of classical waves produced by a double slit (Fischler and Lichtfeldt 1992; Ebison 1975; Haber-Schaim 1975; Feynman et al. 1965). This formalism is subsequently interpreted in probabilistic-statistical terms, applying it to the analysis of microscopic systems, such as the atom, and processes such as interference made with photon beams or low-intensity electrons. The formal/instrumental or axiomatic-formal approach therefore emphasizes the role of the mathematical structure of quantum theory.

This approach is generally proposed through the wave function formulation of quantum mechanics, as illustrated in Table 1 which summarizes the key points of two of the first didactic proposals formulated with this approach (Ebison 1975; Haber-Schaim 1975). At university level the wave formulation was prevalent in first classical books, that still constitute authoritative references (Born 1989; Messiah 1961; Landau and Lifšits 1958). It is a rigorous and coherent approach, but it requires strong skills both on physical concepts and on the mathematical tools to be used, that the use of computer simulations to "visualize" processes (Zollman 2002; Cataloglu and Robinett 2002; Redish et al. 2001; Herrmann 2000; Broklova and Koupil 2007; Mason et al. 2015) can only partially reduce.

Table 1 Key steps of the approaches based on the introduction of the wave function (Ebison 1975; Haber-Schaim 1975)

• Analysis of the photoelectric effect, the radiation pressure or the Compton effect to account for the corpuscular nature of light and to found the relations $E = hc/\lambda$ and $p = h/\lambda$
• Analysis of diffraction and interference patterns, carried out in the laboratory and re-proposed with films on single-photon experiments to recognize the unpredictability of individual impacts even in the regularity of the patterns observed
• To describe the processes, it is assumed that the mathematical formalism to be used is analogous to that with which the interference of classical waves is treated
• It therefore associates an amplitude (a wave function Ψ) to each classical alternative
• The superposition of these wave functions allows to correctly predict the position of the minimums of intensity in the interference pattern
• A statistical meaning is attributed to this association
• The analogy of the interference figures obtained with electrons and neutrons and those obtained with photons, suggests adopting an analogous formal description also for material particles
• The concept of wave packet is introduced as a superposition of plane waves of different frequencies, to attempt to describe a particle
• Finally, the case of a confined particle is analyzed which is described with stationary waves and consequently discrete energy levels
• The quantitative analysis of the Gaussian wave packet makes the uncertainty relations emerge analytically, allowing the measurement process to be discussed
• It follows that it is impossible to describe the trajectory of a particle, since the only acceptable statement is that the particle can be "localized in a finite region of space" at any time
• The formal tools introduced allow to build semi-quantitative models through which: to estimate the atomic dimensions and the energy of the fundamental state of the hydrogen atom; to recognize the inconsistency of Bohr's model; to account for the existence of the meson, of nuclear forces, of the Lorentzian enlargement of spectral lines, of the impossibility of confining an electron to a nucleus

In this type of treatment, the aim is to construct the tools necessary to make the new mechanics understand the potential for interpretation, in particular, in relation to the atomic structure and the phenomena connected to it. It reserves, in proportion, less attention to the recognition of peculiar elements of quantum physics, such as indeterminism, the incompatibility of some quantities. The formalism used has the advantage that, at least in the basic aspects (the use of a function), it is also known to secondary school students. However, every attempt to overcome a first qualitative or semi-qualitative level, even for the simplest aspects, clashes with formal hard difficulties to overcome by high school students. This is essentially the reason why the proposals adopting a formal approach have not had a real use in the school (Stefanel 2008), up until the advent of the computer. A complete, qualified and high level cluster of simulations to support both secondary school and university students' learning of quantum physics was developed by Zollman research group (Zollman et al. 2002).

This formal approach has been transposed into several PER approaches available in literature, especially at the dawn of the interest of didactic research for the teaching of QM (Lawrence 1996; Ebison 1975; Haber-Schaim 1975; Faletič 2015; Giliberti and Marioni 1997). More recently, the use of computers allowed the creation of simulations based on representation of the wave function or probability density able to activate effective learning processes (Zollman 2002; Cataloglu and Robinett 2002; Faletič and Kranic 2019; Passante and Kohnle 2019). Also the approach to the many paths of Feynman (1985) adopts a formal approach, as it provides an axiomatic introduction of the rules for the new mechanics and the exploration of the consequences in the analysis of different phenomena. This proposal was transposed didactically by Taylor (Taylor 1998; Taylor et al. 1998) and subsequently re-proposed without any particular novelty except for an increasingly massive use of computer simulations and documentation of students' learning (Fabri 1996; Borello et al. 2002; Ogborn and Taylor 2005; Hanc and Tuleja 2005; Fanaro et al. 2012; Malgieri et al. 2014).

The approach to quantum fields (Hobson 2005; Giliberti et al. 2004) can be here included. It starts from the assumption that the current view of the microscopic world is based on the quantum field theory and therefore it is important that students face this way of looking at phenomena. Both electromagnetic radiation and matter are proposed, therefore, as fields, emphasizing the fact that electrons and photons are nothing but the quanta of the respective fields (electronic and electromagnetic) and therefore have the same ontological status. The privileged context for this type of approach is the analysis of interference light phenomena and material particle beams (electrons, neutrons, atoms), seen as quanta of the respective fields. This approach promotes a unitary and modern vision of radiation and matter, which is its main value, but requires mathematical skills that are too high to go beyond a first qualitative-descriptive level.

4 Conceptual Approach

A way, alternative to the previous ones, to approach the QM is what has been called "conceptual" here, which focuses on the discussion of the founding elements of the theory starting from the way in which the connection between phenomena and concepts is made. It is based on the choice of a reference phenomenological context, typically two-state system phenomenologies, such as the interaction of photons with polaroid and birefringent crystals to explore the phenomenology of light polarization (Pospiech 1999, 2000, 2003; Michelini 2008; French et al. 1975; Ghirardi et al. 1996; Michelini et al. 2000, 2001; Giliberti et al. 1997), Stern-Gerlach experiments for the exploration of spin (McIntyre 2002, 2012), Mach Zender interferometer (Müller and Wiesner 2002; Kohnle et al. 2015; Pereira et al. 2009) or two slit interferometer (Haber-Schaim 1975; Feynman et al. 1965), two-state potential well (Faletič and Kranic 2019) for analysis of interference phenomena). It discusses the conceptual contents of quantum theory, showing the potential it has in unifying the vision of microscopic phenomena. It develops through the recognition that it is necessary to construct a theory based on the concept of state, no longer tied to intrinsic properties of the system, as in classical physics, but rather to the process by which a system is prepared in that state. The QM, therefore, establishes a new organic way of thinking, which starts from conceptual assumptions that are profoundly different both from classical physics and from the "quantum physics", of which it also incorporates and regains the results (Sakurai 1990; Ghirardi 2004; Michelini 2008; Pospiech et al. 2008; Fischler and Lichtfeldt 1992; Müller and Wiesner 2002).

The central knot of this approach, therefore an indispensable content at every level of education, is the link between the principle of superposition and non-epistemic indeterminism, which characterizes quantum processes. It finds expression in the linear formalism of Hilbert spaces and in particular it appears particularly explicit in the general formulation that Dirac gave of QM (Dirac 1958) and in the most modern formulations of quantum mechanics (Sakurai 1990). The need to abandon a classical description emerges immediately, to adopt a quantum point of view in which the measurement process is characterized as an irreversible projection operation on one of the eigenstates of the observable considered. The simple phenomenological contexts analyzed constitutes the experimental reference to discuss the uncertainty and complementary principles, and the link with measures, the complex relationship between macro-world and micro-world (Mermin 1990; Pospiech 1999, 2000; Ghirardi 2004; Michelini 2008). The representation of the quantum state with a vector ket, an abstract entity unencumbered by any representation, is an aspect that favors the overcoming of the identification of a quantum system with the formal entity that represents its state (Michelini et al. 2000, 2001). In the case of high school level, a treatment using real vectors can be adopted.

Proposals at college level aimed at discussing the concept of state, its vector representation (in complex spaces) and the meaning of the linear quantum principle in the context of interference and entanglement are those of Schneider and La Puma (2002) and Holbrow et al. (2002). The analysis of concepts, in the context of light polarization, is that proposed by Pospiech (1999, 2000, 2003), that focus her approach on the philosophical aspects of quantum theory (Shimony 1989) as cultural contribution for the citizen.

Some words can be said on the choice of the phenomenological context analyzed in the approaches adopting a conceptual approach. The interference phenomena are usually more close to the experiences of students (usually students have some experience on its), but they are not trivial to formalize and for this reason it is privileged by those who adopt a qualitative approach to the quantum concepts, often integrated with the use of computer simulation. Those who analyze light polarization and spin cases can introduce with relative simplicity a formalism accessible to high school students. The advantage of the phenomenology of polarization is the easy reproducibility hands-on experiments (Cobal et al. 2002), the limit is the real spaces of the polarization property is perfectly overlapping to the abstract Hilbert space of state. In the case of spin, this does not occur (McIntyre 2002, 2012). In any case, the treatment of the time evolution of states is not easy to perform analyzing these phenomenologies.

Some recent proposals show how it is possible to introduce the conceptual implication of the time evolution of state analyzing the case of a double potential well or simple quantum systems (Kohnle et al. 2015; Faletič and Kranic 2019; Passante and Kohnle 2019).

5 Research-Based Educational Proposals on QM for School Developed in Italy

Last but not least of this review, it seems useful to share with the international community the results of the research collaboration of Italian PER groups on teaching/learning QM in high school and the impact it has had on teacher education at national scale in the period 2005–2019 (Sperandeo 2004; Battaglia et al. 2011; Francaviglia et al. 2012a, b; Michelini 2010a, b, c; Michelini et al. 2013). The characterizing and valuable elements of this research collaboration have been the cross-referenced development of four different approaches to QM for high school validated both in experiments monitored in schools and in training modules offered to physics teachers of all Italian high schools: Dirac-based approach by means of polarization by Giancarlo Ghirardi, Marisa Michelini and the whole Udine group (Michelini 2008; Michelini et al. 2000, 2001; Ghirardi et al. 1997); approach to the Feynman sum over the paths, developed by Giuseppina Rinaudo in Torino (Borello et al. 2002) and, then, intensively followed by Malgieri et al. (2014); approach to linear formalism using harmonic oscillators developed by Carlo Tarsitani in Roma (Giannelli and Tarsitani 2005); approach based on quantum fields, developed by Marco Giliberti in Milano (Malgieri et al. 2014). The Udine proposal fits the conceptual approach, the other three adopted formal approaches.

From 2006 and till 2019, the Italian PER community was involved in the Scientific Degrees Plan (PLS), promoted by the Ministry of education, research and university (Sperandeo 2004) to motivate young people to study physics. In this plan, the theme of modern physics was chosen by launching the IDIFO project (Didactic Innovation in Physics and Orientation) with the collaboration of 18 educational research groups of Italian universities, coordinated by the University of Udine and under the scientific responsibility of one of us (Michelini et al. 2016). The four different approaches were offered to the different groups to share ideas and each one acts as critical friend for the other to build solid proposals implemented in schools by means of research-based intervention modules and then offered in the IDIFO project. The project IDIFO included actions on a national scale and on a local one. Three main actions were activated at national level: the institution of a II level Master (IDIFO) of 60 cts for professional development of secondary school teachers I in modern physics. Each participant teacher could build their own profile choosing the group of 3 cts courses in an offer of over 150 cts on 5 areas: modern physics, QM or Relativity, with proposals also on statistical mechanics, solid physics, astrophysics, high energy physics; physics in context; RTL and modeling; training orientation; experimentation at school. Five full immersion national summer schools for teachers were carried out in that framework and Quantum Physics was in the program of the Schools as well as a cluster of 10 Experiments for the educational Lab on modern physics.[1] Ten research-based summer schools for talented students of secondary school were carried out on QM, Optical Spectroscopy, Superconductivity and on crucial experiments in Modern Physics. At the local level, educational workshops have been developed co-designed with schools.

Extensive literature documents research outcomes (Battaglia et al. 2011; Francaviglia et al. 2012a, b; Michelini et al. 2013) and offers in Italian language a valuable reference and research contribution to all those interested in educational innovation in modern physics in school (Michelini 2010a, b, c).

6 Conclusion

The different educational proposals on quantum mechanics offer a very wide spectrum of educational proposals and specific paths. They differ for the contents and organization modalities, the choices of the central topics to treat, the vision that is in the background for the quantum concepts. In this review paper, they were organized in three main approaches according to how the basic contents of the QM are organized on a disciplinary level for education: the historical approach, in which the steps that lead to the building of quantum theory are traced; the formal structural approach,

[1] See www.fisica.uniud.it/URDF and in particular http://www.fisica.uniud.it/URDF/laurea/idifo6.htm for the activities and http://www.fisica.uniud.it/URDF/laurea/materiali/index.htm for the material produced (in Italian). The path on QM is in the page http://www.fisica.uniud.it/URDF/secif/mec_q/mq.htm.

in which the adoption of a formalism (usually the wave formalism) is assumed to analyze the consequences of its probabilistic-statistical interpretation; the conceptual approach in which the analysis of the key concepts of the theory are explored usually analyzing the case of two-state systems and introducing the conceptual meaning of the formalism and its role as a conceptual organizer.

The historical approach has the longer tradition in schools and school textbooks (as well as in popular ones). It has generally been proposed as a rational reconstruction of the main interpretative problems of classical physics, the critical discussion of the possible interpretations of phenomena and a reconstruction of crucial experiments of this birth of physics of quanta. It is particularly rich in cultural connections and stimuli to activate a discipline transversal vision of the problems, it is often required in schools for competence development. However, both the mathematical difficulties encountered in seriously addressing this reconstruction, both the limited time available in the school and the conceptual difficulties related to the semi-classical approaches adopted, this approach have led in fact to a narrative and descriptive implementation of phenomena, experiments, hypotheses in which both the scientific value of the research and their historical-cultural value are not discussed or even completely ignored. The physics education research has also shown how many of the students' difficulties in forming a coherent quantum vision of the world are linked to this didactic practice (Zollmann 1999).

Today, qualified and high level complete cluster of simulations can support both secondary school and university students learning to construct a coherent vision of quantum phenomena (Zollman et al. 2002).

The formal structural approach is rigorous and, as has been said, is nowadays often supported with excellent simulations that facilitate the construction of the conceptual interpretation of the formalism adopted. Its main educational value lies in the possibility of visualizing (albeit within certain limits) quantum entities, of drawing on formal structures quite familiar to students such as functions, of favoring analogical reasoning based on phenomenology that students should know, such as that of waves and wave optics. In this approach, the formalism adopted is assumed as an axiom as well as the rules to interpret this formalism. These assumptions are rather appealing for many teachers and presumably for the best students in mathematics, but it emphasizes the adopted formalism, which is often motivated by constructing analogies with classical phenomenology, by means of a reinterpretation in a quantum key and the revision of concepts is loose, shifting students' attention to the computational role of mathematics, rather than its conceptual meaning. The computer relieves students of mere calculation, but it still leaves open the knot of how to give plausibility to the quantum rules underlying the formalism itself.

The conceptual approach focuses precisely on the search for plausibility of quantum rules in the analysis of simpler phenomenology such as those of two-state systems. It aims and brings students directly into the heart of the foundational concepts of quantum theory, with the aim of developing a quantum way of thinking. Typically, a qualitative thought develops which finds its conceptual organization only later in formalism. The perspective is therefore exactly opposite to that adopted in the previous approach. A first limitation of this approach is linked to the lack of those connections and stimuli that the historical approach enhances most. A second limitation, which however also has a formal approach, is to focus the students on a context, without letting them grasp the generality of what that context "teaches".

Studies on conceptual knots and learning difficulties associated with the proposals stimulate a comparison based on the goal to integrate the different perspectives to offer a global framework to the students, as the Italian experience briefly reported here has taught us. The requisite shared is that we cannot offer this topic as the narrative story of a piece of physics, but we have to offer to the students the opportunity to understand and use the basic concepts and formalism.

References

Phys Educ (2000), Special Issues 35(6)

Am J Phys (2002), Special Issues 70(3)

Akarsu B (2011) Instructional designs in quantum physics: a critical review of research. Asian J Appl Sci 4(2):112–118

Baily C, Finkelstein ND (2014) Teaching quantum interpretation. https://Archiv.org/Pdf/1409.8503

Battaglia RO, Michelini M et al (2011) Master IDIFO. In: Hurkett RD, Rogers L (eds) Physics community and cooperation, vol 2. Lulu, Leicester, pp 97–136

Borello L, Cuppari A, Greco M, Rinaudo G, Rovero G (2002) Il metodo della "somma dei cammini" di Feynman per l'introduzione della Meccanica Quantitica: una sperimentazione nella Scuola di Specializzazione per l'Insegnamento. La Fisica Nella Scuola XXXV(2 suppl):125–131

Born M (1953) Atomic physics, VIII ed., Blackie & Son, Glasgow reprint 1989, Dover, New York

Broklova Z, Koupil J (2007) Visualization of hydrogen atom states. In: Ellermajer T, Van den Berg E (eds) Modelling in physics and physics education. Amstel, Girep, Amsterdam, pp 292–302

Brookes JC (2017) Quantum effects in biology: golden rule in enzymes, olfaction, photosynthesis and magnetodetection. Proc R Soc A 473:20160822. https://doi.org/10.1098/rspa.2016.0822

Budde M, Niedderer H, Scott P, Leach J (2002a) The quantum atomic model 'Electronium': a successful teaching tool. Phys Educ 37(3):204. https://doi.org/10.1088/0031-9120/37/3/304

Budde M, Niedderer H, Scott P, Leach J (2002b) Electronium': a quantum atomic teaching model. Phys Educ 37(3):197. https://doi.org/10.1088/0031-9120/37/3/303

Cataloglu E, Robinett RW (2002) Testing the development of student conceptual and visualization understanding in quantum mechanics. Am J Phys 70:238

Cobal M, Corni F, Michelini M, Santi L, Stefanel A (2002) A resource environment to learn optical polarization. In: Proceeding of Girep conference 2002, physics in new fields, Girep, Lund

DM Indicatori Nazionali, http://www.indire.it/lucabas/lkmw_file/licei2010/indicazioni_nuovo_impaginato/_decreto_indicazioni_nazionali.pdf

Dirac PAM (1958) The principles of quantum mechanics. Calderon Press, Oxford

Dür W, Heuslery S (2013) What we can learn about quantum physics from a single qubit. arXiv: 1312.1463v1[physics.ed-ph] 5 Dec 2013

Ebison MG (1975) Introducing the uncertainty principle. In: Loria A, Thomsen P (eds) Seminar on the teaching of physics in schools, vol 2. Girep, Gylendal, pp 220–256

Fabri E (1996) Come introdurre la fisica quantistica nella scuola secondaria superiore. La Fisica nella Scuola XXIX(1-sup):63–80

Faletič S, Kranic T (2019) Combining the two-state system with a matter-wave approach for teaching quantum mechanics in high-school, internal communication

Fanaro M, Otero MR, Arlego M (2012) A proposal to teach light in a unified framework using the Feynman method. In: Problems of education in the 21st century, vol 47, pp 27–39

Faletič S (2015) A mechanical wave system to show waveforms similar to quantum mechanical wavefunctions in a potential. EJP 36:035023

Feynman RP (1985) QED: the strange theory of light and matter. University Press, Princeton

Feynman RP, Leighton RB, Sands M (1965) The Feynman lectures on physics, vol 3. Addison Wesley, Reading

Fischler H, Lichtfeldt M (1992) Modern physics and students' conceptions. Int J Sci Educ 14:181–190

Francaviglia M, Lorenzi MG, Michelini M, Santi L, Stefanel A (2012a) Teachers facing conceptual nodes of QM. http://www.archiv.aplimat.com/2012/Proceedings/New_trends_in_education/Francaviglia_Lorenzi_Michelini_etal_1.pdf

Francaviglia M, Lorenzi MG, Michelini M, Santi L, Stefanel A (2012b) Idifo3—teachers formation on modern physics and mathematical foundations of quantum physics. Aplimat 5:231–240

French AP (1975) Experimental bases for quantum ideas. In: Loria A, Thomsen P (eds) Seminar on the teaching of physics in schools, vol 2. Girep Gylendalm pp 258–272

Ghirardi GC (2004) Sneaking a look at God's Cards. Princeton University Press, Princeton

Ghirardi GC, Grassi R, Michelini M (1996) A fundamental concept in quantum theory. In: Bernardini C, Tarsitani C, Vicentini M (eds) Thinking physics for teaching. Plenum Publishing Corporation Aster, pp 329–332

Ghirardi GC, Grassi R, Michelini M (1997) Introduzione delle idee della fisica quantistica e il ruolo del principio di sovrapposizione lineare. La Fisica nella Scuola XXX(3 Sup. Q7):46–57

Giannelli A, Tarsitani C (2005) A proposal for introducing elementary quantum mechanics at school. In: Proceedings dell'ICPE 2004 Congress 'What Physics should we teach?'. ICPE Durban–South Africa, pp 141–146

Giliberti M, Marioni C (1997) Introduzione di alcuni Elementi di Fisica dei Quanti nella Scuola Secondaria Superiore. La Fisica nella Scuola XXX 3(Supp. Q7):58–67

Giliberti G, Lanz L, Cazzaniga L (2004) Teaching quantum physics to student teachers of S.I.L.S.I.S.-MI. In: Michelini M (ed) Quality development in the teacher education and training. Forum Udine pp 425–429

Greca IM, Freire O Jr (2003) Does an emphasis on the concept of quantum states enhance students' understanding of quantum mechanics? Sci Educ 12(5–6):541–557

Haber-Schaim U (1975) On the teaching of quantum physics in the senior high school. In: Loria A, Thomsen P (eds) Seminar on the teaching of physics in schools, vol 2. Girep, Gylendal, pp 273–284

Hadzidaki P, Kalkanis G, Stavrou D (2000) Quantum mechanics: a systemic component of the modern physics paradigm. Phys Educ 35(6):386–392

Hanc J, Tuleja S (2005) The Feynman quantum mechanics with the help of Java applets and physlets. http://pen-physik.de/w_jodl/MPTL/MPTL10/contributions/hanc/Hanc-Tuleja.pdf

Herrmann F (2000) The Karlsruhe Physics Course. Eur J Phys 21:49–58

Hobson A (2005) Electrons as field quanta: A better way to teach quantum physics in introductory general physics courses. Am J Phys 73(7):630–634

Hoehn JR, Gifford JD, Finkelstein ND (2019) Investigating the dynamics of ontological reasoning across contexts in quantum physics., Phys Rev Phys Educ Res 15:010124

Holbrow Ch, Galvez E, Park M (2002) Photon quantum mechanics and beam splitters. Am J Phys 70(3):260–265

Johansson KE, Nilsson Ch (2000) Experiments in modern physics for the general public. Phys Educ 35(4):256–262

Klassen S (2011) The photoelectric effect: Reconstructing the story for the physics classroom. Sci Educ 20:719

Kohnle A, Baily C, Campbell A, Korolkova N, Paetkau MJ (2015) Enhancing student learning of two-level quantum systems with interactive simulations. Am J Phys 83:560–566

Krijtenburg-Lewerissa K, Pol HJ, Brinkman A, van Joolingen WR (2019) Key topics for quantum mechanics at secondary schools: a Delphi study into expert opinions. Int J Sci Educ 41(3):349–366.https://doi.org/10.1080/09500693.2018.1550273

Landau LD, Lifšits EM (1958) Quantum mechanics, non-relativistic theory. Pergamon Oxford
Lautesse P, Vila Valls A, Ferlin F, Heraud JL, Chabot H (2015) Teaching quantum physics in upper secondary school in France: quantons' versus 'wave–particle' duality, two approaches of the problem of reference. Sci Educ 24:937–955
Lawrence I (1996) Quantum physics in school. Phys Educ 31(5):278–286
Malgieri M, Onorato P, De Ambrosis A (2014) Teaching quantum physics by the sum over paths approach and GeoGebra simulations. Eur J Phys 35:055024
Mason B, Debowska E, Arpornthip T, Girwidz R, Greczylo T, Kohnle A, Melder T, Michelini M, Santi L, Silva J (2015) Report and recommendations on multimedia materials for teaching and learning quantum physics. Il Nuovo Cimento 38 C:105–116
McIntyre DH (2002) Spin and quantum measurement. PH 425 Paradigm 5
McIntyre DH (2012) Quantum mechanics. Pearson Education Inc., San Francisco
Mermin D (1990) Quantum mysteries revisited. Am J Phys 58(8):731–734
Mermin D (1998) What quantum mechanics trying to tell us? Am J Phys 72(3):348–350
Mermin D (2003) From Cbits to Qbits: Teaching Computer Scientist QM. Am J Phys 71(1):23–30
Messiah A (1961) Quantum mechanics, vol I. North-Holland, Amsterdam
Michelini M (2008) Approaching the theory of quantum mechanics: the first steps towards a coherent synthesized interpretation with a supporting formalism. In Jurdana-Sepic R et al (eds) Frontiers of physics education. Zlatni Rijeka, pp 93–101
Michelini M (ed) (2010a) Progetto IDIFO. Fisica Moderna per la Scuola. Materiali, aspetti e proposte per l'innovazione didattica e l'orientamento. MIUR-PLS-UniUD, Udine
Michelini M (ed) (2010b) Formazione a distanza degli insegnanti all'innovazione didattica in fisica moderna e orientamento. MIUR-PLS-UniUD Udine
Michelini M (ed) (2010c) Proposte didattiche sulla fisica moderna, Strumenti per una didattica laboratoriale. MIUR-PLS-UniUD Udine
Michelini M, Stefanel A (2008) Learning paths of high school students in quantum mechanics. In: Jurdana-Sepic R et al (eds) Frontiers of physics education. Zlatni Rijeka, pp 337–343
Michelini M, Ragazzon R, Santi L, Stefanel A (2000) Proposal for QM in secondary school. Phys Educ 35(6):406–410
Michelini M, Ragazzon R, Santi L, Stefanel A (2001) Quantum physics as a way of thinking: an educational proposal. In: Pinto R, Santiago S (eds) PhyTEB 2000. Elsevier Paris, pp 479–482
Michelini M, Santi L, Stefanel A (2013) E-learning in teacher professional development in innovation and formative guidance on modern physics: the case of IDIFO Master's Programs. J e-Learn Knowl Soc 9(2):43–75
Michelini M, Santi L, Stefanel A (2014) Teaching modern physics in secondary school. In: Kajfasz E, Masson T, Triay R (eds) Proceedings FFP14 Marseille 2014, http://pos.sissa.it/archive/conferences/224/231/FFP14_231.pdf
Michelini M, Santi L, Stefanel A (2016) JQM per affrontare nella scuola secondaria i fondamenti di meccanica quantistica. In: Proceedings of Didamatica 2016, http://didamatica2016.uniud.it/proceedings/dati/articoli/paper_96.pdf
Müller R, Wiesner H (2002) Teaching quantum mechanics on an introductory level. Am J Phys 70(30):200–209
Mullis IVS, Martin MO, Foy P, Hooper M (2016) TIMSS Advanced 2015 international results in advanced mathematics and physics. https://timssandpirls.bc.edu/timss2015/international-results/wpcontent/uploads/filebase/full%20pdfs/T15-InternationalResults-in-Mathematics.pdf
Newton RG (2004) What is a state in quantum mechanics? Am J Phys 72(3):348
Niaz M, Klassen S, McMillan B, Metz D (2010) Reconstruction of the history of the photoelectric effect and its implications for general physics textbooks. Sci Educ 94:903
Niedderer H, Deylitz S (1998) Introduction to atomic physics. A concept based on the Schrödinger equation, Textbook for students. Institute of Phys Educ Univ of Bremen Bremen
Ogborn J, Taylor EF (2005) Quantum physics explains Newton's laws of motion. Phys Educ 40(1):26–34

Passante G, Kohnle A (2019) Enhancing student visual understanding of the time evolution of quantum systems. Phys Rev PER 15:010110
Pereira A, Ostermann F, Cavalcanti C (2009) On the use of a virtual Mach-Zehnder interferometer in the teaching of quantum mechanics. Phys Educ 44(3):281–291
Petri J, Niedderer H (1998) Atomic physics in upper secondary school: layers of conceptions in individual cognitive structure. Int J Sci Educ 20:1075–1088
Pospiech G (1999) Teaching EPR paradox at high school? Phys Educ 34:311–316
Pospiech G (2000) Uncertainty and complementarity. Phys Educ 35(6):393–399
Pospiech G (2003) Philosophy and quantum mechanics in science teaching. Sci Educ 12:559
Pospiech G, Michelini M, Stefanel A, Santi L (2008) Central features of quantum theory in physics education. In: Jurdana-Sepic R et al (eds) Frontiers of physics education. Zlatni Rijeka, pp 85–87
Redish EF, Steinberg RN, Wittmann MC (2001) A new model course in applied quantum. Physics https://www.physics.umd.edu/perg/qm/qmcourse/NewModel/index.html
Sakurai JJ (1990) Modern quantum physics, 2nd edn. rev. Benjamin/Cummings, Menlo Park
Schneider MB, La Puma IA (2002) A simple experiment for discussion of quantum interference which-way measurement. Am J Phys 70(3):266–271
Shimony A (1989) Search for a worldview which can accommodate our knowledge of microphysics. In: Cushing JT, Mullin EM (eds) Philosophical consequences of quantum theory. University of Notre Dame Press Paris pp 62–76. https://doi.org/10.1017/CBO9780511621147.003
Singh C (2008) Student understanding of quantum mechanics at the beginning of graduate instruction. Am J Phys 76:277
Sperandeo RM (2004) The pre service physics teacher education model implemented by the F.F.C. research project involving & Italian universities: guidelines and preliminary results. In: Michelini M (ed) Quality development in the teacher education and training. Forum Udine pp 89–96
Stadermann HKE, van den Berg E, Goedhart MJ (2019) Analysis of secondary school quantum physics curricula of 15 different countries. Phys Rev Phys Educ Res 15:010130
Stefanel A (2008) Impostazioni e percorsi per l'insegnamento della meccanica quantistica nella scuola. G Fis 49(1):15–33
Styer DF, Balkin MS, Becker KM, Burns MR, Dudley CE, Forth ST, Gaumer JS, Kramer MA, Oertel DC, Park LH, Rinkoski MT, Smith CT, Wotherspoon TD (2002) Nine formulations of quantum mechanics. Am J Phys 70:288–297
Taylor EF (1998) Demystifying Quantum Mechanics. http://www.artsaxis.com/eftaylor/
Taylor EF, Volkov S, O'Meara JM, Thornber NS (1998) Teaching Feynman's sum-overpaths quantum theory. Comput Phys 12(2):190–198
Toraldo Di Francia G (1975) Teaching formal quantum mechanics. In: Loria A, Thomsen P (eds) Seminar on the teaching of physics in schools, vol 2. Girep, Gyldendal, pp 318–329
Viau J, Moro L, Tintori Ferrerira MA (2011) Física Moderna En La Escuela Secundaria: La Narrativa En El Aula. In: Porta L, Zelmira A, Sarasa C, Bazan S (eds) VI Jornadas Nacionales sobre la Formaci6n del Profesorado. Universidad National, Mar del Plata, p 229
Wan T, Emigh PJ, Shaffer PS (2019) Investigating how students relate inner products and quantum probabilities. Phys Rev Phys Educ Res, 15:010117
Weissman EY, Merzel A, Katz N, Galili I (2019) Teaching quantum mechanics in high-school—Discipline-Culture approach. J Phys: Conf Ser 1287:012003
Zollmann D (ed) Research on teaching and learning quantum mechanics. Papers presented at the Annual meetings NARST, Published at www.phys.ksu.edu/perg/papers/narst/ (1999)
Zollman D (2002) The formal reasoning of quantum mechanics: can we make it concrete? Should we? In: Michelini M, Cobal M (eds) Developing formal thinking. Forum, Udine, pp 85–93
Zollman D, Rebello SN, Hogg K (2002) Quantum mechanics for everyone: hands-on activities integrate with technology. Am J Phys 70(3):252–259

Quantum Cryptography as an Approach for Teaching Quantum Physics

Gesche Pospiech

Abstract Quantum physics is not only a fundamental physical theory but recently it promises big technological advances in the so-called 2nd quantum revolution, especially in quantum information. These advances rely on the basic research done since about 1980 while looking for a clarification of the fundamentals of quantum physics. The corresponding experiments and insights paved the way for quantum cryptography which is the first development in quantum information leaving the laboratory and going into practice. These developments should have its influence on teaching physics at school level in order to give students insight into a fascinating and fundamental part of modern physics. Herewith they can experience the fundamental notions of quantum physics: non-determinism, superposition and uncertainty. To achieve this goal an approach exploiting recent results of quantum cryptography and combining it intimately with the fundamentals seems promising. In addition such an approach also permits to introduce students with help of Dirac notation to mathematical structures of quantum physics. This might additionally support understanding quantum concepts without retreating to vague metaphors or descriptions. Therefore, teaching quantum information, especially quantum cryptography, at school may serve for motivating students and at the same time impart insight into physics research and the nature of physics. We present a corresponding teaching–learning proposal that was conducted with teacher students.

Keywords Quantum cryptography · Physics education · High school teaching

1 Introduction

Quantum physics as a fundamental physical theory forms an important part of physics education at school where students should experience modern physics and its implications. A main goal of education in quantum physics is the insight into its differences from classical physics, thus helping in building a modern physics worldview.

G. Pospiech (✉)
TU Dresden, 01067 Dresden, Germany
e-mail: gesche.pospiech@tu-dresden.de

B. Jarosievitz and C. Sükösd (eds.), *Teaching-Learning Contemporary Physics*, Challenges in Physics Education,
https://doi.org/10.1007/978-3-030-78720-2_2

Hence it also contributes greatly to general education. In popular media and in some textbooks, quantum physics is "mystified" to some extent, being called "strange" or stating that "the behavior of electrons cannot be understood". These wordings take not sufficiently into account that since the invention of the Bell inequalities (Bell 1964) or the experiments of Aspect (Aspect et al. 1982) great advances in clarification of the meaning of quantum physical notions such as entanglement or uncertainty have been made. Research on these properties gained strong impetus by the invention of algorithms especially suitable for quantum computers (Deutsch 1985; Shor 1994). Recently these beginnings resulted in fast progress such that the possibility of a true and usable quantum computer seems on the edge (Acín et al. 2018). In October 2019, Google reached quantum supremacy for the first time. In this context teaching quantum information means that the development of quantum physics during the twentieth century is being followed and can be brought to school. Whereas in the first half of the century the first generation of quantum technology was developed—among many other examples—in inventing and building the laser, using the photoelectric effects in LED or solar cells or finding the transistor, toward the end of the century quantum technology of the second generation accelerated the research quite efficiently: the laser was used for manipulating quantum systems, transistors were miniaturized and other technologies help in promoting the 2nd quantum revolution, namely quantum simulation, quantum metrology, quantum sensors, quantum information and quantum communication. These developments in the last two fields were again and again covered by popular media showing the interest and fascination of a wide public (Metz and Zhong 2020; https://www.sciencealert.com/quantum-computers). Especially research in quantum cryptography or the building of a quantum computer and its possible influences on economy and technology are being highlighted and critically discussed. Considering that those reports sometimes are quite exaggerated it is difficult for the average reader to judge the real impact of the reported research. This aspect even increases the relevance of learning quantum physics as a part of general education. As future citizens also students not preparing to pursue a scientific career should be able to judge the plausibility of such reports and in this way to take part in the societal discourse. In order to be able to do so a basic understanding of the principles of quantum physics is necessary. However, there are hints (Singh 2008) that traditional teaching approaches to quantum physics do not generally reach an appropriate understanding of its core concepts, e.g., uncertainty. In addition, often quantum physics is considered too complicated because of its mathematical formalism.

But recently, two-state systems come increasingly into focus as a suitable means for introducing and describing the central notions of quantum physics such as superposition, uncertainty, entanglement and the measuring process. Some studies give hints that these two-state systems are especially suitable for demonstrating the differences of quantum and classical physics and that on the whole also students find this approach easier than the traditional one (Sadaghiani 2016). The reason might be that the two-state systems can be described much easier from a mathematical point of view (finite dimensional Hilbert space, discrete eigenvalues, etc.) and hence allow for focusing on the concepts instead of elaborate mathematics. From the conceptual point of view, they are not so intimately related to classical notions of position and trajectory and therefore allow better to grasp the quantum thinking. So two-state systems can be viewed as kind of toy system that is easy to handle and suitable for learning the basics of quantum physics.

Independent of these advantages two-state systems may serve for describing many different quantum phenomena. The mainly used example is the polarization of photons as a model for its spin. But also other applications are important in the research on quantum information and in quantum technology as, e.g., the ground state and excited state of atoms or ions, Josephson contacts and many others. So a teaching approach with a focus on two-state systems has not only physical and mathematical advantages but also covers topics where physicists are actually working on thus increasing the insight into actual research.

From these different perspectives, quantum cryptography is an as well motivating as accessible topic. Therefore, already twenty years ago, it was proposed to use quantum cryptography or quantum computing as an approach for teaching quantum physics on high school level, because this would allow to reach directly the core of quantum physics and hence also cover some world view aspects (Pospiech 1999, 2000a). Since then quantum cryptography has made big advances even recognized by the public (e.g., Deutsch 1985). So the question arises of how to teach quantum physics in the context of quantum cryptography without neglecting the goal of imparting the quantum concepts as part of general education. Moreover, the important objective of giving the students insight into the heart of quantum physics and the differences between classical and quantum physics, including philosophical aspects and providing a modern physical worldview, can be satisfied. Herewith it has to be taken into account that even if two-state systems are relatively easy to handle it remains an abstract topic which cannot be adequately taught without a basic amount of formalization.

Therefore, teaching this current topic requires careful alignment of theoretical aspects and visualization by different representations, by suitable well-chosen metaphors (Brookes and Etkina 2007; Pospiech 2019) or model experiments. So we start with the fundamental concept of superposition which is central and builds the basis for all other quantum physical phenomena. The uncertainty is especially crucial for quantum cryptography to ensure safe encryption whereas the properties of entanglement might be best demonstrated by quantum teleportation. In the quantum computer all these quantum concepts mesh. The characteristics of the measuring process (with non-determinism) are important for all these applications and also for forming the physical worldview.

2 Quantum Information in Textbooks and Teacher Students' Views

Teaching quantum physics is strongly dominated by traditional approaches still often used in textbooks on school level and on university level. Also the curricula of different countries can be summarized to a kind of "core curriculum" which is dominated by an approach via double slit, atomic physics and the wave function (Stadermann et al. 2019). In order to develop a modern appropriate teaching–learning-sequence incorporating quantum information aspects the perspective of teacher students and educators as well as available textbooks have to be taken into account.

2.1 Quantum Information in Textbooks

Textbooks play a twofold role in teaching: On one hand, they have to be consistent with the official curriculum, on the other hand they have to be accepted by teachers. Also often teachers rely on textbooks for shaping their lessons. Therefore we analyzed textbooks designed for the last two years of school before the final exam (allowing the entrance into university) with respect to the covered topics of quantum physics. All the chosen text books appeared after 2012. As a rough overview, we give the number of pages for the different topics which provides a first insight into the focus of the book (see Table 1).

We see that traditional topics such as photo effect and Compton effect play an important role, also because they often can be part of the written final exam. Wave-particle duality is being replaced increasingly by the notion of quantum object. Aspects concerning measuring process or interpretation only play a minor role, with one exception. Only one book explicitly treats the EPR-experiment. However, entanglement with the motivating application of teleportation occurs in several books. From this, we conclude that the recent quantum information approaches slowly find their way into the physics textbooks even on school level.

2.2 Acceptance by Teacher Students

The question arises of how teacher students see the possibility of including aspects of quantum information into their teaching. So teacher students were asked about which topics they think as suitable for teaching at school and which contents were not suitable.

The study was performed by an online questionnaire sent to universities throughout Germany who offer study courses for future physics teachers. This questionnaire aimed at revealing the expectations of students as well as their experiences

Table 1 Overview of the distribution and amount of topics in quantum physics covered in the different text books for school

Topic	Book 1	Book 2	Book 3	Book 4	Book 5	Book 6	Book 7
History	2	–	–	–	–	–	–
Photo effect	2	4	–	4 + 2	5	3	–
Compton effect	1	–	–	2	2	1/2	–
Wave-particle duality	5	QO: 2	QO: 3	QO: 2 + 4	1/2	4	1+ QO: 1
Uncertainty	4	2	2	3	3	2	1
Double slit/interference	2	2 + 2 + 2	1	2	4 + 4	1	1 + 2
Measuring process	3	1	1/2	1	implicit	-	1
Entanglement	1	1	1, teleportation: 2	implicit		-	1, teleportation: 3
Interpretation	3	1	1/2	3	2 incl. EPR	1	1/2

The numbers mean the number of pages reserved for the corresponding topic

with the study specifically in quantum physics (Schöne 2020). It was intended for students from their third year on when they had attended the lectures on quantum theory in a theoretical physics course. A special focus was on the question if they felt prepared for teaching quantum physics at school and which topics they saw as adequate for school teaching. This was not a representative study but its results were quite in line with the expectations and therefore might serve as hints to what teacher students think about their possibilities concerning future teaching. Overall 110 students took part in this survey.

The results show that students mostly orient themselves toward the contents they themselves learned at school and are quite skeptical toward new contents or very mathematical aspects (Schöne 2020). The results are presented in Table 2. The students seem to have a clear opinion about the "teachability" of some of the mentioned concepts. This can be seen from the consistent answers. This is the case for example for uncertainty, matter-wave-dualism, interference, laser, photo effect and Compton effect. These topics are usually taught at school and also covered in university lectures. So it is not astonishing that students see these topics as "teachable". On the other hand, they do not see as "teachable" the topic Schrödinger equation, which is highly mathematical and includes partial differential equations, or the phenomenon of Bose-Einstein condensates which are neither taught at school nor normally included in basic lectures at university. With other topics the students seem to be undecided or mostly show no clear opinion. Among those topics the measuring process as well as entanglement, quantum cryptography and quantum computer are found. This we interpret in a way that the teacher students perhaps would like to treat

Table 2 In the table, it is indicated which quantum concepts or notions are considered as suitable or not suitable for teaching at school by the teacher students

Percentage	Suitable	Not suitable
>60%	Uncertainty, laser, photo effect, Compton effect, matter-wave-dualism, interference	Schrödinger equation, Bose–Einstein-condensates
	Measuring process, quantum numbers, entanglement	Quantum computer, quantum cryptography, measuring process, probability function, radiation formula
<30%	Angular momentum, spin, Bose–Einstein condensates, quantum computer, quantum cryptography, Schrödinger equation, potential well, probability function, radiation formula	Uncertainty, angular momentum, spin, laser, photo effect, Compton effect, entanglement, matter-wave-dualism, interference, quantum numbers

The left column indicates the percentage of students explicitly saying the concepts are either suitable or not suitable. The concepts printed in underlined are answered quite consistently and clearly. Concerning the others, the answers are not unambiguous

it in school but are not sure if they can achieve this, how to do it and if the school students would be able to grasp it. Therefore we decided that a seminar introducing them to these topics would be helpful.

In planning the seminar we also took into account that we could identify two groups of students with a different focus (via cluster analysis with complete linkage method). Most students (around 75%) could be attributed to the so-called "broad type". These students could imagine to teach applications such as laser, Compton and photo effect as well as quantum computer or quantum cryptography. They are also aware of the concepts of uncertainty/entanglement/measuring process. The other smaller group of about 25% of students is more "traditionally oriented" and would stick to topics like interference or quantum numbers in the context of atomic physics (Schöne 2020).

Besides these perceptions, the teacher students expressed their general needs toward an additional seminar oriented toward aspects of teaching quantum physics. These included teaching methods, knowledge of students' conceptions, applications of quantum physics and deepening the insight into the concepts of the interpretation of quantum theory without going too deeply into the formalism (Schöne 2020). In the construction of the seminar also these points were taken into account.

3 Construction of Course (Design Principles)

The described twofold goal of giving students insight into the heart of quantum physics together with the differences between classical and quantum physics as well as into current applications requires careful alignment of theoretical aspects, experimental realization and visualization by different representations. Therefore, different from many other proposals, we decided not first to teach quantum physics and then treat quantum cryptography as an application but to show the intimate connection between the principles of quantum physics, how they are exploited in quantum cryptography and why they make it a secure method of communication in a coherent teaching–learning sequence.

In the sense of educational reconstruction, we first clarify the content matter before we construct the course.

3.1 *Analysis of Relevant Terms and Concepts*

The goal of the seminar is to bring together the aspects of mathematics, concepts of quantum physics and their application in quantum cryptography and physics education. The interlinking should contribute to an understanding of the quantum concepts in a deep and natural way. Therefore, in the following, we discuss the most relevant concepts, namely quantum state, superposition, quantum behavior and uncertainty. The entanglement we will only mention, because it is not unavoidable. As described above the teaching–learning sequence should be founded on two-state systems.

3.1.1 Mathematical Constructs Describing Quantum State and Superposition.

Understanding the concept of quantum state is the crucial point in learning quantum physics. The basis lies in the construct of Hilbert space where the quantum states are represented as abstract vectors. From the vector space structure, the strictly valid superposition principle follows as an immediate consequence. Students have to learn that any given state (vector) can be represented with respect to different bases of the vector space and that they can switch between them without changing the state itself. With respect to polarization this means that a state could be represented, e.g., in the basis of the σ_x or the σ_z operator. This is crucial for understanding the Q-bit as smallest unit in quantum information. For supporting understanding of superposition verbal representations such as "superposition of possibilities" or the famous Schrödinger metaphor "catalogue of possibilities" could be used. With this we expect that students see no difficulty in imagining that several possibilities can exist at the same time.

3.1.2 Mathematical Constructs Implying Quantum Behavior

Quantum behavior is characterized by two aspects: one is the change of quantum states by operators, the second is the measuring process. The operators carry all the information of a quantum system: Their eigenvalues are possible measuring results and the eigenvectors represent those states which upon measurement arrive with certainty at the corresponding eigenvalue as a result. All this leads to stochastic behavior of quantum objects in the measuring process. In the context of quantum cryptography, this is a central precondition for the generation of a perfect random key necessary for secure encrypting.

3.1.3 Mathematical Construct Interpreted as Uncertainty

In quantum physics, the functions describing a classical physical system are replaced by operators. Correspondingly, the Poisson brackets become commutators. If these are different from zero (because of non commutativity) quantum phenomena (e.g. non-compatibility of physical quantities) occur. The consequences imply that some physical quantities cannot have well defined values at the same time. This leads to uncertainty, which in combination with the non-determinism is indispensable for security of quantum cryptography. The mathematical structures of linearity and Hilbert space as state space also imply that the state space of a combination of several quantum objects is represented as a tensor product. The resulting "entanglement" is quite different from the direct product of vector spaces used in classical physics, leading to the notion of "non separability", a concept not imaginable in classical physics. This concept is used, e.g., in the EK91 protocol of quantum cryptography but it is not necessary for securing privacy.

3.2 Teaching Principles

A motivating teaching–learning sequence should generate in students the "need to know". This can successfully be reached by problem-oriented teaching and relevance to everyday life. In the last years, topics with relation to quantum information in the broadest sense, above all reports on quantum computing or quantum cryptography with respect to privacy gain public attention. This finds its expression in online articles, in blogs and in classical newspapers. The ability of critical reviewing popular media and judge their adequateness as well as asking appropriate questions contributes to the general education. It is also a central goal of school to enable students to take part in societal discourse. Therefore anchored instruction, e.g., use of news articles seems appropriate. Such an approach (with other topics) has been developed and successfully evaluated (Kuhn and Müller 2014). Hence the teaching unit on quantum cryptography could start with a suitable recent article from a newspaper or similar media (see, e.g., Metz and Zhong 2020; https://www.sciencealert.com/quantum-computers). This opens the opportunity for learners to articulate their opinions and questions. This phase is important to unfold the topic not only with respect to physics questions but also to possible implications for societal life. In addition, the teacher as well as the learners can draw on possible pre-knowledge and pre-conceptions.

Besides the relation to everyday life also the physical significance should find its expression. For this, not only qualitative descriptions should be followed but also the mathematization. Lessons on quantum physics may serve for learning also about the nature of physics from several perspectives. These are not only the philosophical aspects or the many applications with its interplay of technology and physics insights, but also the use of mathematics. The students should learn how (selected) results with physics meaning can be gained only with mathematics, so that its structural role besides technical calculations can become clear. The gap between a qualitative explanation and the mathematical description should be bridged by use of multiple representations which, taken together, form a visualization of the different aspects of the relation between the mathematical and the physical world (Ainsworth 2008; Geyer and Kuske-Janßen 2019).

The teaching–learning sequence as presented here restricts itself to the absolutely necessary contents and neglects all other aspects, e.g., technical realization even if this might also be interesting to some students. This restriction was made because the available time at school is limited. But as described above the focus is twofold and therefore might have some emphasis on worldview, hence it could easily be unfolded toward more philosophical aspects.

4 Description of Teaching–Learning Sequence

According to the teaching principles referred to above, the teaching–learning sequence starts by presenting a recent news article on quantum cryptography inviting the students to ask questions. These questions will be structured and embedded into a problem-based methodology. After the teaching unit, the students should be able to judge the news article with respect to correctness or if exaggeration can be recognized.

As it cannot be supposed that students are familiar with the techniques of cryptography in a first step they are introduced to some basics about classical cryptography they need to know. This part restricts itself to the most simple *encrypting procedure*, the use of a so-called one-time-pad. The most important point is that the students recognize that the key used for encryption has to be absolutely random and to be kept secret. The students should also understand that every message can be represented by a sequence of 0 and 1. These facts lead to the statement that the main tasks to be solved are the generation of a suitable (perfectly random) key and its confidential exchange among the sender of a message (Alice) and its receiver (Bob).

After this preface we introduce in a step by step manner—always in strong connection with procedures of quantum cryptography—the main quantum principles: quantum state, superposition, quantum behavior, uncertainty and—if time and the learning group allows for it—entanglement (see Fig. 1).

The teaching sequence uses throughout the existence of single photons. In order to show the relevant properties of photons their introduction is supported by simulations or screen experiments (see www.quantumlab.de). A crucial point then is the description of the polarization states of photons. This first requires knowledge of

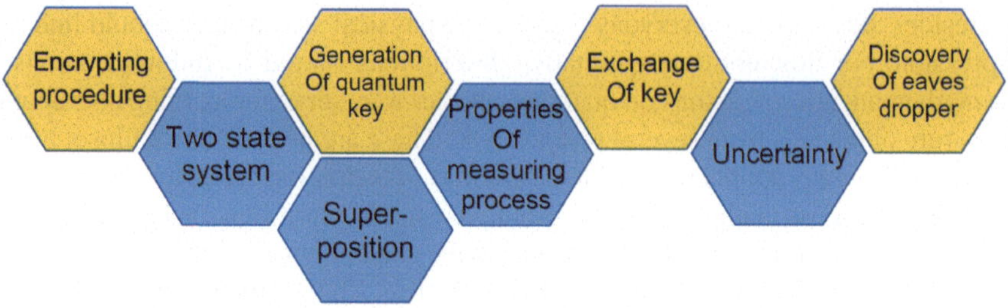

Fig. 1 The construction of the course showing the interlacing of quantum concepts and their application in quantum cryptography

the property "polarization" and an understanding of the concept of state in quantum physics and its representations in different forms. An anchor for introducing this concept is provided by model experiments using polarizers and calcite crystals and exploring the interaction of light with these means (Pospiech 2000b; Michelini et al. 2005). From these experiments the notion of two-state systems is introduced, using a visualization with different representations connecting it to the spin. Those representations are iconic and symbolic, preparing for the increasing mathematization. In this way, it is derived that the physics of photons allows using them as a basis for an encrypting procedure by suitably attributing 0 or 1 to the basis states.

The next task to be solved is the *generation of a quantum key*. This needs the *superposition principle* together with the *properties of the measuring process*, foremost its non-determinism. The resulting randomness of measuring results serves to establish the required properties of the key: a perfect random sequence of 0 and 1. At this point it might be suitable to explicitly remind the students of the classical procedure where normally the random number generators only are called random but are calculated, hence possess an inner structure. This difference contributes to recognizing the different nature of classical and quantum physics. As in this part of the course the foundations are introduced sufficient time should be taken to explain, to discuss and to apply the knowledge. Geometrical elements can be used as representation of the two-state system and prepare the following use of mathematics and the Dirac notation. Herewith it is important to connect every mathematical or symbolic element with its physical meaning and the relevance for the problem setting from the beginning. The insights of this step are generalized to quantum properties such as non-determinism of measuring results, representing the measuring process by a projection and changing the state during measuring process.

In the next step the *quantum key has to be exchanged* between Alice and Bob. This exchange heavily relies on the properties of states, eigenstates and the measuring process. The students have to gain the insight that an eigenstate is reproduced in the corresponding measurement, even if the measuring result of a superposed state cannot be predicted. In order to arrive at this insight different bases for the description of a state of a quantum system are introduced and supported by a model experiment with several polarizers. Also here suitable visualization is used and the formalized description with the help of the Dirac notation is being introduced (Pospiech 2000b).

A central part of the course is the confirmation that *a spy could be detected* with help of the BB84 protocol. This protocol requires the *quantum uncertainty*. Several steps can be taken to create insight into this phenomenon. The principle can be clarified with model experiments (see Pospiech 2000b) which imitate the thought experiment of putting several rotated Stern-Gerlach apparatus in series. The students could predict which outcome they would expect in these experiments and then see that it is different, creating a cognitive conflict. In a next step, this insight can be deepened and made more concrete by formal calculations using the Dirac notation. Then the BB84 protocol with detecting a spy is treated with help of worksheets. Here the students can experience how the uncertainty is working. This part serves also as a strong indication of the differences between classical and quantum physics.

As an addendum serving for deepening and giving an outlook the Non Cloning Theorem can be treated. If the BB84 protocol is analyzed carefully one could come to the conclusion that the eavesdropper just could catch the quantum state of the photon coming from Alice, copy it before sending it further to Bob and then await the exchange of bases and results between the two. Here the No Cloning Theorem says that this copying process is not possible with a quantum copying machine. The proof in an elementary version is not complicated and can be shown with help of the Dirac Notation and a few simple rules.

Another extension could be done in using the EK91 Protocol where not single photons are being sent from Alice to Bob, but an EPR-Pair is sent from a source in between, one part to Alice, the other to Bob. The remaining part of the protocol is similar to the BB84. However it could be an opportunity to introduce entanglement and show it in action. The entanglement is also important in that it represents the basis of the working of efficient quantum computers. But this part is not needed for a basic version of the course.

In the end, the problem of a secure communication seems to be solved and the news article can be discussed on a thorough knowledge base.

5 Outlook

In teaching quantum physics several approaches are possible with very different focus and mathematical requirements. The most traditional approach is quite complicated from a mathematical and conceptual viewpoint in using atomic physics and treating the position-momentum uncertainty as an important content. The proposed approach inspired by quantum optics showed that incorporating quantum cryptography into a teaching–learning sequence on quantum physics based on two-state systems is working and fulfills its goal of imparting the basic quantum notions such as non-determinism, superposition and uncertainty. It requires less complicated mathematics and leads directly to the heart of the interpretation debate.

This approach has been tested in a seminar for teacher students complementing the standard lecture on quantum theory which is obligatory for all future physics teachers. During this seminar the teacher students gained content knowledge as well as pedagogical content knowledge. The detailed results of the evaluation are presented elsewhere (Schöne 2020).

However, in transforming such a seminar for school students additional aspects have to be considered. Above all special attention has to be paid to the fact that the school students do not have the mathematical background and the mathematical-physical experience of teacher students. Therefore suitable elementarization and visualization have to be used. The visualization does include multiple representations with different degrees of abstraction. Such representations of a pictorial, geometrical, symbolic or algebraic character have to be linked to each other in order to provide anchors for understanding and thinking tools for developing qualitative argumentation, conceptual understanding and quantitative predictions. The representations and their interlinking should also support the disciplinary discourse in class induced by teachers. Therefore high school teachers should know and master the different representations (including Dirac notation) and to be aware of their implications, their abstractness and their advantages and disadvantages.

References

Acín A et al (2018) The quantum technologies roadmap: a European community view. New J Phys 20:080201

Ainsworth S (2008) The educational value of multiple-representations when learning complex scientific concepts. In: Gilbert JK, Reiner M, Nakhleh M (eds) Visualization: theory and practice in science education. Springer Netherlands, pp 191–208. https://doi.org/10.1007/978-1-4020-5267-5_9

Aspect A, Grangier P, Roger G (1982) Experimental realization of Einstein–Podolsky–Rosen–Gedanken experiment: a new violation of Bells inequalities. Phys Rev Lett 49:91–94

Bell JS (1964) On the Einstein-Podolsky-Rosen Paradox. Physics 1:195–197

Brookes DT, Etkina E (2007) Using conceptual metaphor and functional grammar to explore how language used in physics affects student learning. Phys Rev Spec Top-Phys Educ Res 3:010105

Deutsch D (1985) Quantum theory, the Church-Turing principle and the universal quantum computer. Proc R Soc London A400:97–117

Geyer M-A, Kuske-Janßen W (2019) Mathematical representations in physics lessons. In: Pospiech G, Michelini M, Eylon B-S (eds) Mathematics in physics education. Springer International Publishing, pp 75–102

https://www.sciencealert.com/quantum-computers. Last accessed 2020/12/27

Kuhn J, Müller A (2014) Context-based science education by newspaper story problems: a study on motivation and learning effects. Perspect Sci 2:5–21

Metz C, Zhong R (2020) The race is on to protect data from the next leap in computers. And China Has the Lead. The New York Times. https://www.nytimes.com/2018/12/03/technology/quantum-encryption.html. Last accessed 2020/12/27

Michelini M, Ragazzon R, Santi L, Stefanel A (2005) Discussion of a didactic proposal on quantum mechanics with secondary school students. Il Nuovo Cimento C 27:555–567

Pospiech G (1999) Quantenkryptographie. Ein elementarer Zugang zur Quantentheorie. Physik in der Schule 37:201–205

Pospiech G (2000a) Quantencomputer - Was verbirgt sich dahinter? Der Mathematische Und Naturwissenschaftliche Unterricht 53:196–202

Pospiech G (2000b) Uncertainty and complementarity: the heart of quantum physics. Phys Educ 35:393

Pospiech G (2019) Pre-service teacher's views on the use of metaphors for describing the concepts of uncertainty and entanglement in teaching quantum physics. Int J Phys Chem Educ 11:1–5

Sadaghiani HR (2016) Spin first vs. position first instructional approaches to teaching introductory quantum mechanics. In: Jones DL, Ding L, Traxler A (eds) 2016 Physics education research conference proceedings. Compadre, pp 292–295

Schöne M (2020) Die Verbesserung der Lehramtsausbildung in der Quantentheorie https://nbn-resolving.org/urn:nbn:de:bsz:14-qucosa2-334882. Last accessed 2020/12/27

Shor PW (1994) Algorithms for quantum computation: discrete logarithms and factoring. In: Proceedings 35th annual symposium on foundations of computer science. IEEE, pp 124–134

Singh C (2008) Assessing student expertise in introductory physics with isomorphic problems. I. Performance on nonintuitive problem pair from introductory physics. Phys Rev Spec Top Phys Educ Res 4:010104

Stadermann HKE, van den Berg E, Goedhart MJ (2019) Analysis of secondary school quantum physics curricula of 15 different countries: different perspectives on a challenging topic. Phys Rev Phys Educ Res 15

www.quantumlab.de. Last accessed 2020/12/27

Milq—Quantum Physics in Secondary School

Rainer Müller and Oxana Mishina

Abstract The milq approach to quantum physics for high schools focuses on the conceptual questions of quantum physics. Students should be given the opportunity to engage with the world view of modern physics. The aim is to achieve a conceptually clear formulation of quantum physics with a minimum of formulas. In order to provide students with verbal tools, they can use in discussions and argumentations we formulated four "reasoning tools." They help to facilitate qualitative discussions of quantum physics, allow students to predict quantum mechanical effects, and help to avoid learning difficulties. They form a "beginners' axiomatic system" for quantum physics.

Keywords Quantum physics · Physics education · Reasoning tools

1 Introduction

Without doubt, quantum physics is one of the more difficult areas for teaching at high school level. Why is this so? When trying to pinpoint the difficulties more precisely, one finds that the explanation is not easy: (1) Quantum physics is experimentally poorly accessible with high school resources? Yes it is, but that is probably not the real problem. (2) Quantum physics is so difficult to teach because it is a mystery and nobody understands it? Popular accounts often give the impression that essential parts of quantum physics are conceptually poorly understood, especially in those areas (such as the measurement process) that are particularly interesting for discussion at high school. We will show in this article that, contrary to this belief, it is possible to discuss even the most difficult aspects of quantum physics at high school level if only the appropriate terms and concepts are used. (3) It is difficult to teach quantum physics

R. Müller (✉) · O. Mishina
TU Braunschweig, Bienroder Weg 82, 38106 Braunschweig, Germany
e-mail: rainer.mueller@tu-bs.de

O. Mishina
Dipartimento di Fisica, Università di Trieste, Via Valerio 2, 34127 Trieste, Italy

B. Jarosievitz and C. Sükösd (eds.), *Teaching-Learning Contemporary Physics*, Challenges in Physics Education,
https://doi.org/10.1007/978-3-030-78720-2_3

at high school level because it is mathematically so complicated. This is not correct either. Consider the case of electricity: The Maxwell equations are mathematically much more complicated than the Schrödinger equation. Nevertheless, in the case of electricity, it has been possible to formulate the content in such a way that the basics of electricity are routinely taught in high school today.

The comparison between quantum physics and electricity is helpful to understand in which direction quantum physics education could be successfully developed. In electricity, it has been possible to identify some basic features for teaching which are regarded as "essential" for understanding. The corresponding terms have become so much a part of our everyday life that we take them for granted: Nobody would doubt that a proper introduction to electricity would include the concepts of charges and static electricity, current and voltage, stationary circuits, magnetic effects, and induction and electromagnetic waves. In quantum physics, such a clarification process about basic and essential concepts has apparently never taken place. The curriculum is often based on the history of discovery—an approach that certainly would not work well in the case of electricity.

The aim of the milq concept has been to provide such a clarification of basic concepts in quantum physics for high school. The aim was not to make small corrections to existing concepts, but to restructure the subject matter for teaching from a different perspective. The German version of the course has been online since 2001 and is continuously developing. Since 2018, an English version is available, and it will continue to grow (milq.info/en/).

In Germany, quantum physics has been an established part of the schools physics curriculum for several decades. There is a considerable amount of experience and research on teaching and learning quantum physics in secondary schools. The first curriculum explicitly based on the investigation of students' conceptions and learning difficulties was that of Fischler and Lichtfeldt (1994). They aimed at a "minimal conception" that strongly emphasizes the difference between classical and quantum phenomena. References to classical physics (Bohr's atomic model) are largely avoided. Already earlier, Brachner and Fichtner (1977, 1980) had proposed a curriculum which, following the "Feynman Lectures" (Feynman et al. 1966), focuses on the concept of probability amplitude. The "quantum mechanical fundamental principle" (see below) is explained using the example of the Mach–Zehnder interferometer. Several curricula have been developed based on Feynman's "arrow formalism" (Feynman 1985), e.g., the proposals of Bader (1994), Küblbeck (1997) or Werner (2000). In the English-speaking world, the "Visual Quantum Mechanics" concept by Zollman et al. (2002) can best be compared with these curricula in terms of their physical and mathematical requirement level. Simulation programs are used here, for example, to examine the spectra of LEDs and trace them back to the band structure in solids. Also the approach by Michelini et al. (2000), where the polarization degrees of photons are related to states and operators, addresses to a similar audience.

2 The Milq Concept

The milq approach has been developed since the mid-1990s in Munich and later in Braunschweig (Müller 2003; Müller and Wiesner 2002). It focuses on the conceptual questions of quantum physics. Students should be given the opportunity to engage with the world view of modern physics. It is not intended to strive for continuity with the notions of classical physics. Instead, especially those aspects of quantum physics are emphasized that involve a radical break with the classical concepts. The most prominent examples are Born's probability interpretation, which implies the departure from classical determinism, and superposition states in which quantum objects do not possess classically well-defined properties like position, trajectory, or energy.

Aspects of proper language play a major role in the milq approach, because many misconceptions and learning difficulties in quantum physics are related to language. Our language has developed in dealing with the phenomena of classical physics. It is only partially suitable for describing the completely different world of quantum phenomena. In university physics, the mathematical formalism is the medium for communication about quantum physics. Language can play a subordinate role, because the clarifying reference to formalism is always possible in case of doubt. Linguistic inaccuracies, laboratory jargon, and simplifying abbreviations are commonplace among physicists without causing much damage. This is different in high school physics. Here we do not have access to the mathematical formalism so that we have to rely on language to communicate about quantum physics. For this reason, great emphasis was placed on the conceptual side of quantum physics: Conceptual clarity was the main focus in the development of the milq approach. It was considered important to provide concise terms that help us to speak systematically about quantum phenomena. One example of a helpful notion is the term "preparation," which describes the state of quantum objects operationally and in qualitative terms. It is thus the equivalent of the wave function in mathematical formalism, because the wave function is used to mathematically describe quantum objects that have undergone a certain preparation process.

3 Reasoning Tools

Within the milq approach, we formulate a set of four qualitative rules that are supposed to contain the basic traits of quantum physics. They are called reasoning tools (Küblbeck and Müller 2002). They can serve as a support in qualitative discussions about quantum phenomena. They enable students to predict certain kinds of quantum behavior and help to avoid learning difficulties. In this sense, the reasoning tools may be regarded as a kind of "qualitative mini-axiomatic" for quantum physics. The four rules are as follows:

Rule 1: Statistical behavior
Single events are not predictable, they are random. Only statistical predictions (for many repetitions) are possible in quantum physics.

Rule 2: Interference of single quantum objects
Interference occurs if there are two or more "paths" leading to the same experimental result. Even if these alternatives are mutually exclusive in classical physics, none of them will be "realized" in a classical sense.

Rule 3: Unique measurement results
Even if quantum objects in a superposition state need not have a fixed value of the measured quantity, one always finds a unique result upon measurement.

Rule 4: Complementarity
Exemplary formulations are: "Which-path information and interference pattern are mutually exclusive" or "Quantum objects cannot be prepared for position and momentum simultaneously."

At the time when the reasoning tools were formulated, the concept of entanglement seemed too far away from high school physics to be taken into account. From today's perspective, entanglement appears to be an essential feature of quantum physics, especially under the increasingly important aspect of quantum information. If future applications in quantum technology will yield more experience what can be "done" with entanglement, it should possibly be added. One should add, however, that even in the mathematical theory of quantum mechanics, entanglement is not an independent axiom but a consequence of the principle of superposition.

4 The Physical Background of the Reasoning Tools

4.1 Rule 1: The Probabilistic Nature of Quantum Physics

Rule 1 contains Born's probability interpretation in qualitative form. The probabilistic character of quantum physics may be regarded as the one essential difference to classical physics. Classical physics is deterministic. If you throw a basketball at the right speed and angle, you can be sure that it will hit the basket—and this is repeatable. On the contrary, in the double-slit experiment with single electrons, the location where the next electron is detected cannot be predicted—nor can it be controlled experimentally. This is a general rule: In quantum physics, individual events cannot generally be predicted. This inability to predict is not based on subjective ignorance. The experiments on Bell's inequality show that there are no local hidden variables that could govern the outcome of an experiment. According to quantum physics, there is true randomness in nature.

However, quantum physics is not completely without laws. Its laws are statistical in nature. Many repetitions of the same experiment result in a distribution of measured values that—except for statistical deviations—is predictable and reproducible. In addition, there are certain states (eigenstates of observables) in which the outcome of experiments can be predicted with a probability of 1. The success of quantum physics is based on these regularities, making it the best tested theory in physics.

4.2 Rule 2: Interference and Superposition States

Rule 2 is closely linked to a rule that was emphasized by Feynman et al. (1966). It was called the "fundamental principle" by Brachner and Fichtner (1977). It reads as follows: "If there are different possibilities (paths) for the occurrence of a particular event and if the experimental setup does not determine uniquely that only one specific possibility was chosen, interference always occurs. If, on the other hand, each event in the experimental setup leaves a certain trace that can be used to decide which of the various possibilities was chosen, then interference will never occur."

A paradigmatic application of the fundamental principle is the quantum mechanical double-slit experiment. If, with a suitable measuring device, it can be decided through which of the slits an electron has passed, no interference occurs. If, on the other hand, there is no means for assigning a specific slit to the electrons, they form an interference pattern on the screen.

The double-slit experiment can also be used to illustrate the statement: "In a superposition state, none of the conceivable alternatives will be realized in a classical sense." Within the double-slit apparatus, the electrons are in a quantum mechanical superposition state in which the property "position" cannot be attributed to them. The electrons do not go "either left or right." This assumption would be incompatible with the experimental observation of double-slit interference. Instead, they are in a superposition of both alternatives. Neither of the two classical alternatives (left or right) is actually realized. In quantum physics, electrons can no longer be regarded as objects with a definite position or trajectory. This fundamental insight has important consequences for our concept of atoms where it is crucial that electrons are delocalized objects.

4.3 Rule 3: The Measurement Postulate

Rule 3 contains the measurement postulate of quantum mechanics. Even with quantum objects in superposition states, measurements yield definite results. The result of a measurement is always one of the eigenvalues of the measured observable.

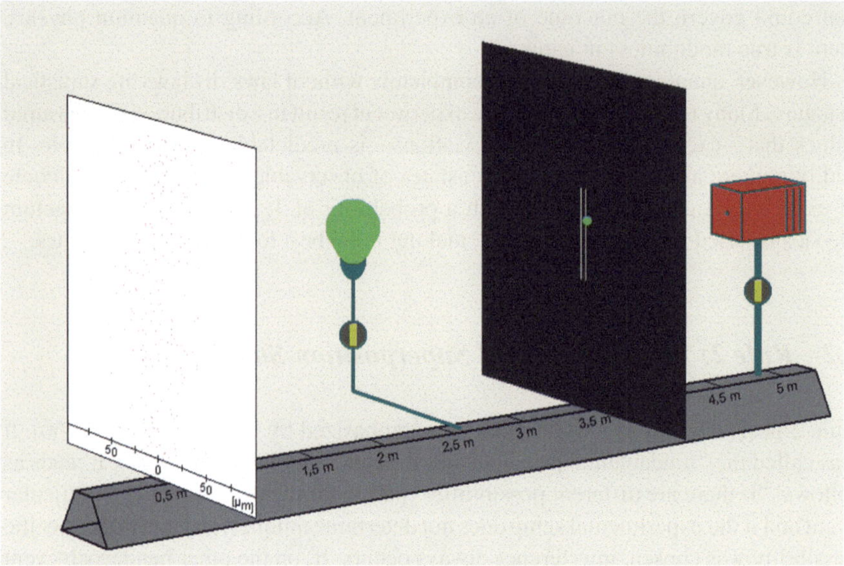

Fig. 1 Screenshot of the milq double-slit simulation software

Following Feynman et al. (1966), this can be illustrated in the double-slit experiment by placing a light source behind the slits to illuminate the electrons passing by (Fig. 1). The electrons scatter some of the light, and the scattered light is registered by a detector or the eye of an observer. In this way, a measurement of the observable "position" is performed. It turns out that for each individual electron, a flash of light is seen at a specific position behind one of the slits (visible in Fig. 1). This means: In the position measurement, each electron is found at a well-defined position—even if the electrons were in a state before the measurement in which they did not have the property "position." A similar position measurement is made at the detection screen: Each electron is found at a well-defined spot on the screen, even if it was in a delocalized state before the detection.

4.4 Rule 4: Complementarity

The Feynman lamp can also be used to illustrate rule 4. When the lamp is switched on and registers the path of each electron, no double-slit interference pattern emerges on the screen. Instead, a distribution without internal structure appears. Which-path information and interference pattern cannot be realized simultaneously, but are mutually exclusive. This is an example of two complementary quantities, such as location and momentum in Heisenberg's uncertainty relation.

4.5 *Epistemological Status of the Reasoning Tools*

One advantage of the reasoning tools is particularly noteworthy: they are strictly correct. Nothing that is stated in the four rules has ever to be taken back, even if one advances to university level and to the most complicated facts of quantum mechanics. This distinguishes them from many models of school physics or chemistry, which have to be "unlearned" in the course of further studies and replaced by successor models (for example in the field of atomic physics).

The reasoning tools also claim to contain the qualitative key aspects of quantum physics quite completely. Of course, they lack anything that can be only expressed within the framework of mathematical formalism. In an epistemological sense, however, nothing entirely new is added, no matter how deep one goes.

5 Key Experiments

In the milq approach, the reasoning tools are introduced by means of key experiments in which they become particularly evident. The aim is to give students the opportunity to test the application of the rules in real-life situations. Two key experiments are discussed in detail. Simulation software has been written for both experiments. It is provided for free at the milq web page milq.info:

1. The first key experiment is the double-slit experiment with single quantum objects, which according to Feynman et al. (1966) "has in it the heart of quantum mechanics." The simulation program for the double-slit experiment provides an interactive environment for exploring its features in a variety of situations, including the measurement process with the Feynman light source.
2. The second key experiment is the experiment of Grangier et al. (1986), which is illustrated in Fig. 2. Single photons are investigated alternatively at a single beam splitter or in a Mach–Zehnder interferometer with two beam splitters. The experiment consists of two parts: (a) If beam splitter 2 (shown in Fig. 2 with dashed lines) is not present, the photons encounter a single beam splitter. After passing it, they are registered by two detectors. The detectors measure the coincidence of single photons that are reflected or transmitted at beam splitter 1. (b) If beam splitter 2 is present, the experiment is a Mach–Zehnder interferometer where each photon can reach a detector via two paths.

6 Application of the Reasoning Tools: Beam Splitter and Mach–Zehnder Interferometer

Let us illustrate the application of the reasoning tools with the second key experiment:

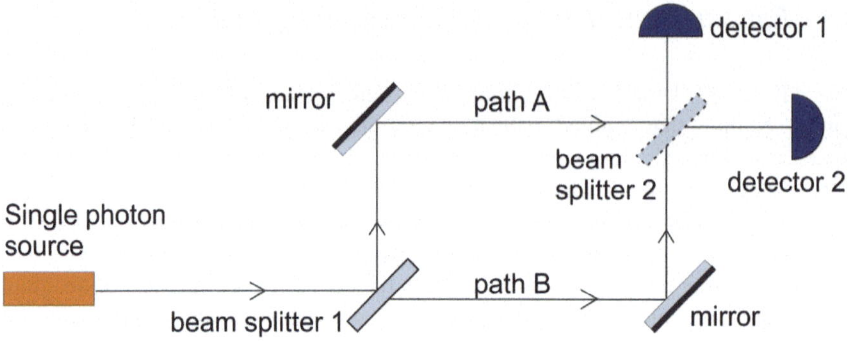

Fig. 2 Schematic setup of the experiment by Grangier et al. (1986). In the milq simulation software, the two detectors are replaced by detection screens, and the beam is widened to obtain a spatial interference pattern

1. **Rule 3**: Without beam splitter 2, the two detectors 1 and 2 perform a position or path measurement. According to rule 3, each position measurement has a definite result. Exactly one of the two detectors clicks, never both. The two detectors show perfect anticoincidence. On measurement, a single photon is always found at a definite position, never at two positions at the same time. The experiment by Grangier, Roger, and Aspect received considerable attention because this anticoincidence provides genuine evidence for the quantum nature of light. The result cannot easily be explained by semiclassical models.
2. **Rule 1**: There is no means to predict at which of the two detectors the next photon will be detected. No physical feature exists that determines whether a photon is transmitted or reflected at the beam splitter. However, a statistical prediction can be made: If the experiment is repeated very often, about half of the photons are found at detector 1 and half at detector 2.

If the second beam splitter is present, rule 1 can be applied too: when a photon hits the screen, its energy is released locally at a specific spot at which position the next spot is detected cannot be predicted. However, the distribution that is formed when many spots are detected is reproducible: it is the interference pattern known from classical optics (Fig. 3).

3. **Rule 2**: If beam splitter 2 is inserted, there are two classical alternatives for the experimental result "detector 1 clicks": a photon may have gotten there via path A or via path B. According to rule 2, interference occurs when the path length of the two interferometer arms is varied. This has indeed been observed in the experiment of Grangier, Roger, and Aspect. The simulation program is based on a somewhat different mechanism for interference. It is assumed that the two paths are slightly different in length, and the beam is widened by a lens. This leads to an interference pattern with a ring-like structure, as shown in Fig. 3.
4. **Rule 4**: To demonstrate rule 4, the experiment shown in Fig. 3 can be extended by polarization filters in both arms. Depending on their relative orientation, they

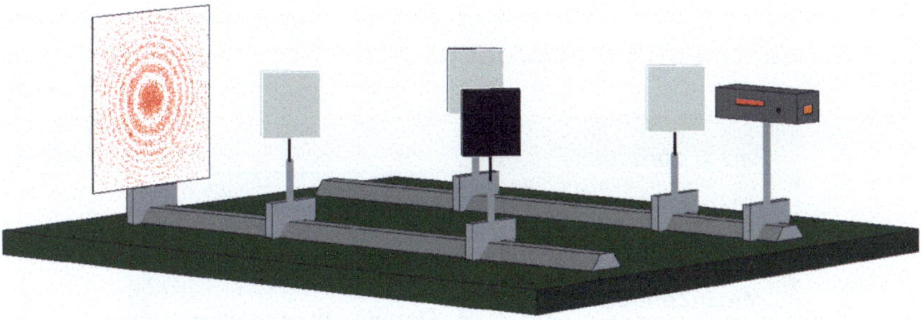

Fig. 3 Interference pattern known from classical optics emerges from the detection traces of many single photons. Screenshot of the milq Mach–Zehnder simulation program

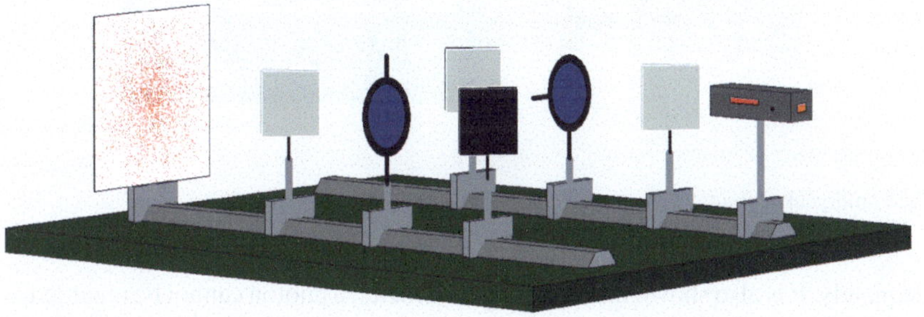

Fig. 4 Which-path information is encoded with polarization filters. No interference pattern appears

imprint path information on the photons through polarization. If both polarization filters are parallel, no which-path information is encoded into the polarization degree of freedom, and the interference pattern is found on the screen. However, with polarization filters set perpendicular to each other, which-path information is encoded and no interference pattern appears (Fig. 4). Which-path information and interference pattern are mutually exclusive—an example of complementary quantities.

7 Curriculum

The detailed curriculum of the milq course has already been discussed in Müller and Wiesner (2002). We will, therefore, only give a brief outline. The course consists of two main parts which form a spiral curriculum (Fig. 5). First, the quantum physical behavior of photons is examined, using the Mach–Zehnder simulation software. It is shown that light may show both wave and particle aspects in the same experiment. However, none of our classical models is sufficient to describe the phenomena

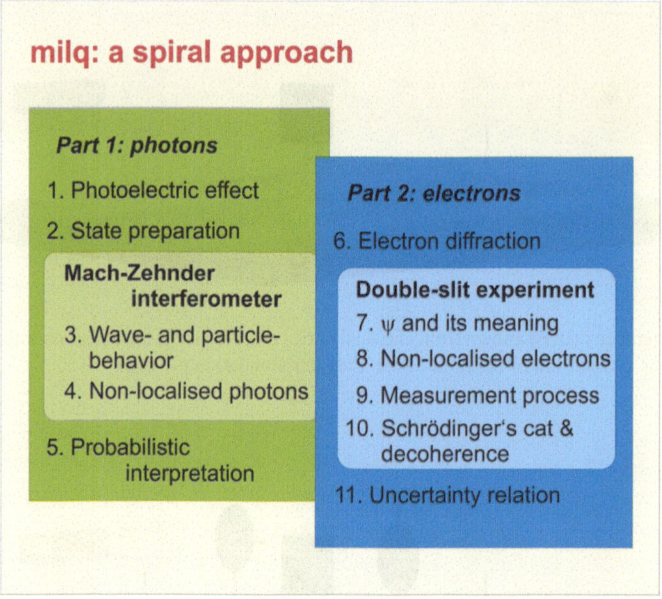

Fig. 5 milq curriculum

adequately. It is also shown that in an interferometer, a photon cannot be regarded as a particle-like entity which is localized in one of the interferometer arms. Born's probability interpretation is necessary to explain the experimental results. In the second part, electrons are studied (mainly in the context of the double-slit experiment). The same topics as with photons are treated, but on a conceptually higher level. The probability interpretation is formulated with the wave function. Still the discussion remains qualitative, i.e., the wave function is only considered as an abstract entity. Finally, some quite advanced topics are discussed: the quantum mechanical measurement process, the paradox of Schrödinger's cat, and its resolution by decoherence.

8 Empirical Results

The milq course has been evaluated at the time of its conception. The results are fully documented in Müller (2003) (in German); some of the results have been published already in Müller and Wiesner (2002). Here we only give a brief overview. In the evaluation, a mix of qualitative and quantitative methods was used. To assess whether students at high school level are able to deal with the challenging subject matter, the following studies were conducted with high school students in grade 13 (more precisely, "high school" means German Gymnasium which is visited by about half the students of a cohort, one third of them chooses physics):

- teaching experiments with individual students ($N = 8$),
- semi-structured interviews on student conceptions ($N = 23$),
- questionnaire on student conceptions ($N = 60$),
- results of written and oral exams.

The questionnaire on students conception contained questions with open answers as well as statements that the students had to rate on a five-point Likert scale from 1 (strongly agree) to 5 (strongly disagree). Four indices were formed from the total of 29 items in the questionnaire, reflecting the students' conceptions on the structure of the atom, determinism/indeterminism, quantum mechanical properties, and the uncertainty relation. In addition, an overall index was calculated by averaging over all 29 items. Both the overall index and the sub-indices were scaled so that the value +100 stands for completely adequate quantum mechanical concepts, while −100 means completely inadequate concepts. The results are given in Table 1. For the high school students, the mean value of the overall index was + 55.8 with a standard deviation of 19.5. Such a high value can be interpreted as an indication that the students have successfully built up the desired quantum mechanical concepts through the milq course.

It was difficult to define a suitable control group because the content of the milq course was far removed from the standard high school curriculum. We decided to compare the high school students of the experimental group with a group consisting of 35 first-year physics students from the University of Munich which had not yet attended quantum physics courses in university. They formed a "group of learners with good knowledge of physics" where at least below-average performance was not to be expected. The conceptions of this group were compared with those of the experimental group. The mean value of the overall conception index in this comparison group was +35.2, i.e., 20.6 points lower than in the test group. The difference is statistically highly significant ($p < 0.1\%$). The effect size d is 0.97; it is

Table 1 Evaluation results for questionnaire on students' conceptions questionnaire

Student conceptions on	Experimental group (high school students)	Control group (university students)	Effect size d
Structure of the atom (6 items)	+60.9	+40.8	0.65**
Determinism/indeterminism (9 items)	+51.6	+37.4	0.47*
Dynamical properties of quantum objects (3 items)	+71.6	+41.9	0.83***
Uncertainty relation (10 items)	+51.5	+30.2	0.92***
Overall index (29 items)	+55.8	+35.2	0.97***

Effect size is Cohen's d; asterisks denote the level of significance (*$p < 0.05$; **$p < 0.01$; ***$p < 0.001$)

therefore a large effect. The differences in the individual areas are also consistently significant to highly significant with medium to large effect sizes (Table 1).

9 Application on a European Scale

Until 2018, the teaching material for the milq course was only available in German. This made it difficult to use the milq concept outside the German-speaking world. In 2018, a first attempt was made to promote the milq approach in Italy. The material was used as basis for teacher in-service training at the University of Trieste. Italy is currently in the process of changing the school curricula in physics by adding modern physics. Teachers are being offered structured training to prepare them for the new teaching (Michelini et al. 2013). A number of teachers approached the University of Trieste and asked for an ad hoc seminar to adapt their knowledge to their teaching needs. The topics of the seminars were chosen based on the requests of the participants after they were introduced to the four quantum reasoning rules. The following sequence of seminars has emerged:

1. Introduction of the "quantum reasoning tools,"
2. the photoelectric effect,
3. single photon in a Mach–Zehnder interferometer,
4. measurement of Planck's constant using LED,
5. construction of a spectrometer used to measure a Planck's constant with LEDs.

Although there was no systematic evaluation, the feedback from the teachers was positive. They found the work with the reasoning tools useful and were interested in continuing the training. The activity will be continued and published in the near future.

References

Bader F (1994) Physik in der Schule 32:250
Brachner A, Fichtner R (1977) Quantenmechanik für Lehrer und Studenten. Schroedel, Hannover
Brachner A, Fichtner R (1980) Quantenmechanik. Schroedel, Hannover
Feynman RP (1985) QED: the strange theory of light and matter. Princeton University Press, Princeton
Feynman RP, Leighton RB, Sands M (1966) The Feynman lectures on physics, vol III. Addison-Wesley, Reading
Fischler H, Lichtfeldt M (1994) Physik in der Schule 32:276
Grangier P, Roger G, Aspect A (1986) Europhys Lett 1:173
Küblbeck J (1997) Modellbildung in der Physik. Landesinstitut für Erziehung und Unterricht, Stuttgart
Küblbeck J, Müller R (2002) Die Wesenszüge der Quantenphysik – Modelle, Bilder, Experimente. Aulis, Köln

Michelini M, Ragazzon R, Santi L, Stefanel A (2000) Proposal for quantum physics in secondary school. Phys Educ 35(6):406–410
Michelini M, Santi L, Stefanel A (2013) J e-Learn Knowl Soc 9. https://doi.org/10.20368/1971-8829/833
Müller R (2003) Quantenphysik in der Schule. Logos, Berlin
Müller R, Wiesner H (2002) Am J Phys 70:200
Werner J (2000) Vom Licht zum Atom. Logos, Berlin
Zollman DA, Rebello NS, Hogg K (2002) Am J Phys 70:252

The Role of Mathematics in Teaching Quantum Physics at High School

Gesche Pospiech, Avraham Merzel, Giacomo Zuccarini, Efi Weissman, Nadav Katz, Igal Galili, Lorenzo Santi, and Marisa Michelini

Abstract As quantum physics (QP) requires substantial mathematical knowledge, a thorough elementarization by reduction of complexity or with help of visualizations is needed without neglecting the goal of deeper understanding. First, we present how the visualization of the Dirac formalism could contribute to a deeper understanding of the central concepts of QP. Then, we present two teaching/learning proposals (TLP) for high school students based on the Dirac notation. Both approaches show how suitable strategies can improve student understanding of topics rooted in math and indicate significant success of the students in understanding, interpreting and implementation of the basic concepts of QP. The first approach presented here uses the concept of "wavity" or "wave-particle-duality" in the context of Dirac notation. The second approach emphasizes the measurement process and, on this basis, moves on to discuss the quantum state. It is asked whether and how it is possible to use the formal structure of these concepts to support students in interpreting them. In an inquiry-based teaching–learning sequence on high school level, educational strategies for consolidating the grasp of quantum concepts through math were developed and implemented in teaching experiments conducted in the framework of design-based research. Results of data analysis in both cases show that the adopted strategies are largely successful.

Keywords Quantum physics · Mathematics in physics · High school education

G. Pospiech (✉)
TU Dresden, 01062 Dresden, Germany
e-mail: gesche.pospiech@tu-dresden.de

A. Merzel · E. Weissman · N. Katz · I. Galili
The Hebrew University of Jerusalem, 9190501 Jerusalem, Israel

G. Zuccarini · L. Santi · M. Michelini
Research Unit in Physics Education, University of Udine, 33100 Udine, Italy

B. Jarosievitz and C. Sükösd (eds.), *Teaching-Learning Contemporary Physics*, Challenges in Physics Education,
https://doi.org/10.1007/978-3-030-78720-2_4

1 Introduction

In this paper, we bring together different approaches to teaching quantum physics (QP) at high school level, in order to make the corresponding mathematical structures understandable for high school students and to use them for furthering its understanding. We regard this as important for several reasons:

- A fundamental reason: Mathematization is inherent in doing physics. Therefore, learning the role of mathematics in physics, including its quantitative aspects, contributes to learning about the nature of physics. In addition, students should experience that mathematical means can provide additional or more precise insights and that they allow us to make predictions (Galili 2019).
- Specifics of QP: Mathematics supports the physical intuition of physicists by the emergence of structures with a meaning in the physical world. The fact that laws are found violating basic assumptions of classical physics, e.g., the non-existence of trajectories, thus implying the loss of determinism, shows the power of a mathematical description and the difficulties which are inherent in its physical interpretation. So, in QP, the mathematical structure is crucial.
- Motivation: Treating modern physics in (high) school lessons is quite important for the motivation of students. Among the areas of modern physics, especially the theories of relativity and QP capture the interest of many students and also contribute to an adequate physical world view.

But, as physics theories of the twentieth century are to a great extent determined by mathematical structures, they require advanced mathematical abilities. So, special efforts are needed to teach these topics at school adequately. Therefore, overarching this joint paper is the fact that QP, as a fundamental physical theory, should form an important part of physics education at school as well as at college or university.

In the last thirty years, a growing number of research-based teaching/learning proposals (TLPs) and instructional tools concerning QP have been developed at secondary school level. A recent comprehensive review of this literature (Krijtenburg-Lewerissa et al. 2017) evidenced that many difficulties students experience are related to the difficulty to connect quantum behavior to the physical reality as they see it, which results in a mix-up of classical and quantum concepts. Based on the available literature, the authors of the review conclude that a non-mathematical, conceptual approach to the content can lead this student population to adequate understanding. However, they highlight the scarcity of data on difficulties with content they classify as "complex quantum behavior", i.e., entities and processes relying on the quantum state: the concept itself and its formal representation, including superposition, collapse and time dependence. On the other hand, while an abstract mathematical treatment of physical systems can represent a hindrance to effective learning, the paradigm shift from the classical picture to the quantum one imposes limitations and constraints to qualitative descriptions. Firstly, the classical lexicon is largely deficient and misleading when applied to quantum concepts. Consider, for instance, the notions of "wave" (Sect. 3.4.2) and of "state" (Sect. 4.3). Secondly, fundamental entities present an in-built mathematical component: In order to interpret the basic meaning of the quantum state, students need to make sense of probability distributions of measurement results. Lastly, quantum systems cannot be visualized directly, only their formal representations can be (at least to a certain extent).

However, the recent advances in quantum information open new perspectives. The corresponding recent shift from the atomic quantum theory focusing on wave functions to quantum optics with photon spin or polarization states brought with it completely new approaches to teaching. The related focus on two-state systems with its much easier structures, e.g., the small (mostly only two or four dimensional) Hilbert space and the direct simple calculation of eigenvalues and eigenvectors allows us to emphasize better the concepts of QP and its differences from classical physics. In addition, the two-state system may serve as a model for many different systems, among them polarization of light, electron spin, ground and excited state of an atom and many more. Therefore, the question arises how to facilitate the learning of the corresponding physics, mathematics and their interplay, because even if the mathematics is relatively easy for two-state systems, it still requires a certain level of abstraction. Therefore, the contributing groups developed strategies that help in grasping the mathematical structures and its meaning for QP. These strategies comprise conscious and systematic visualization of the formalism, active learning methods with focus on concepts and inquiry-based learning.

In general, as the quantum model is distant from daily experience and from other models encountered by students so far, students need a strong support in the interpretation of quantum behavior. Mathematical sense-making in physics might represent a natural basis for facilitating the interpretation of qualitative descriptions. This process should rely on the transition between the different representations available, in order to shed light onto the formal entities at hand. Given the relative lack of data on learning difficulties with topics rooted in the mathematical structure of QP, a need emerges to explore whether and how it is possible to turn this structure from a hindrance for learning into a support for the consolidation of concepts.

2 Interpretation and Visualization of Mathematics in Quantum Physics[1]

Using mathematical means in QP for understanding on the level of high school needs a suitable elementarization including a conscious transition between different representations.

[1] This section is contributed by the PER group number 1.

2.1 Relevance of Representations and Transforming Between Them

Generally, physical processes, laws or relationships can be represented in different ways. However, one single representation mostly only shows a certain aspect of a physical process or object. Therefore, it seems clear that only the interplay of multiple representations can give a complete picture of a given content (Ainsworth 1999; Geyer and Kuske-Janßen 2019; Lemke 1998). So, students should experience the "disciplinary affordances" of physics by using different representations and their interrelation. Being able to read, to manipulate and to relate representations is part of gaining physics knowledge and doing physics and belongs to the "way of knowing" in physics (Airey and Linder 2009). Mutually complementing representations allow for appropriate meaning-making. On the other hand, relating different representations to each other requires skill and might lead to cognitive overload (Geyer and Kuske-Janßen 2019; Ainsworth 2008). So, the use of representations has to be carefully balanced. Especially in the context of abstract mathematical structures which are crucial in describing QP concrete representations might support the interpretation of these structures in the light of quantum concepts and thus further the understanding.

By the term "visualization", we understand the conscious use and combination of multiple representations in order to shed light onto mathematical structures from the physics perspective. The representation of mathematical structures includes the use of intuitive graphical or geometrical tools and the mapping of the physical processes into formal structures or vice versa. It is also possible to use experimental realizations, together with simulations or animations (pictorial elements) or to connect the physical–mathematical description to concrete pictures of processes. A central possibility is to visualize the mathematical structures directly by means of suitable graphical, symbolic or algebraic representations in order to allow for building mental models and thus to better clarify the nature of quantum concepts.

Educators should be aware of the existence of different representations and their interplay together with their meanings and possible interpretations. They should be able to combine representations in order to create a visualization of the quantum physical meanings of a mathematical structure for their students.

2.2 Mathematics Forces a Physical Interpretation

The process of the physical interpretation of mathematical structures generally includes several steps, which are related to each other. In the first step, the physical objects or processes are mapped "literally" onto mathematical elements, in its most simple way just formally relating a physical quantity to a symbol in a formula. The second, more complex, step concerns the relation of the mathematical operations and their physical meaning (Ainsworth 1999). In the third step, these meanings will be framed by a general conception of the physical theory, which again will retroact to the meanings (Ainsworth 2008). So, the conceptual framework of a theory determines the meaning of mathematical structures and vice versa. This representational and meaning-making process is especially difficult in the context of QP because of its complex, abstract and paradoxical features.

Therefore, the kind of representations has to be chosen very carefully. They should mirror the mathematical formalism as it is crucial for putting the quantum constructs on a firm basis, highlight the strong interrelation between mathematical constructs and quantum concepts and might even have a conceptual role. We will highlight this in the following with the example of two-state systems.

2.3 *Representations of Two-State Systems and Its Mathematical Structure*

In order to visualize the two-state systems, representations with different grades of abstractness can be used. There are representations with the help of (model) experiments or thought experiments. Such a thought experiment could be the Stern–Gerlach apparatus (see Fig. 1a). The quantum behavior might also be represented with model experiments using polarizers or calcite crystals which can easily be realized in class (Fig. 1b). The corresponding experiments are especially suitable for demonstrating the uncertainty (Pospiech 1999, 2003). On a more abstract level, a pictorial representation, e.g., showing in a striking manner the two basis states could be used (Fig. 1c). In an additional step of abstraction, this representation can be modified in order to prepare formalization with help of the Dirac notation (Fig. 1d). The mathematical properties can then be described with the Bloch circle (Fig. 1e) which is a more refined representation and requires mathematical argumentation and

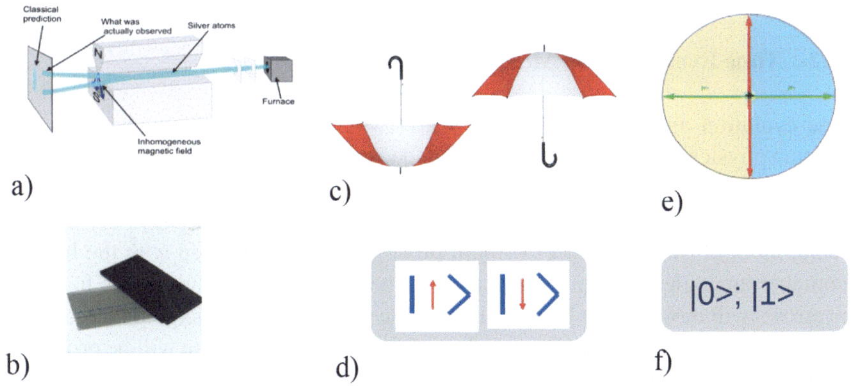

Fig. 1 Different representations for two-state systems. **a** experimental (Stern–Gerlach apparatus), **b** model experiment (polarizers), **c** pictorial (showing up and down), **d** symbolic (Dirac notation with symbols), **e** graphical (Bloch circle as simplification of Bloch sphere), **f** algebraic (Dirac notation)

knowledge (Pospiech 2003). The use of Bloch circle instead of the Bloch sphere allows for avoiding the complex numbers. The most abstract representation is the algebraic one with the full Dirac notation, together with its calculation rules (Fig. 1f). All these different representations can be set into relation to each other and used in parallel, amounting to a double or even multiple representation.

In the light of the advantages of several representations for the same object, we now show how the different types shown in Fig. 1 can be used to support understanding and handling of two-state systems and the basic quantum concepts such as superposition, uncertainty and entanglement. Herewith, it is necessary to clarify which mathematical or physical aspect is highlighted in each of them and how they relate to each other in order to support learning. Furthermore, we will discuss in which way a visualization combining representations with different degree of abstractness could contribute to a deeper understanding of quantum phenomena.

2.3.1 Superposition

The mathematical concept of superposition of vectors in a vector space can be visualized in different ways. In general, vectors and their superposition are represented by arrows, a geometrical representation. In quantum physics, the mathematics takes on a different meaning (see Sect. 4, Table 1). The possible states of a two-state system can be represented by arrows with the caveat that these could be misunderstood as spatial vectors. The physical behavior is then taken into account by the representation in the Bloch sphere (or simplified: the Bloch circle), a graphical representation. This can be combined with a symbolic (Fig. 2a) or algebraic representation (Fig. 2b). This combination might support the meaning-making of the superposition principle for quantum systems, of the symbolic Dirac notation and prepare its algebraic use by stimulating the visual imagination of mathematical elements.

2.3.2 Time Evolution and Measuring Process

Time evolution and measuring process are two inherently different processes in quantum physics excluding each other. An adequate picture could be the Bloch circle combined with Dirac notation [see Figs. 1e and 2, Dür and Heusler (2016)] so that we have a graphical–symbolic visualization addressing different ways of building mental models. The differences of both processes can be clarified with the help of geometric analogies, e.g., animated arrows: In case of time evolution, the heads of the arrows will move on the Bloch sphere or the circle without changing their length. In the case of the measuring process, the arrow representing a state will be projected onto one of the eigenstates.

Table 1 Comparison between different forms of vector superposition

	Superposition of forces	Quantum superposition: decomposition of the state vector
Visualization		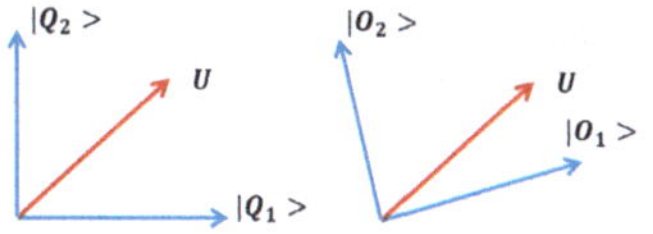
Represented notion	A physical quantity: net force	No physical quantity: a quantum state
Number of physical entities involved	More than one force acting on a system	One entity, the state vector of the system, that is decomposed in a given basis (in the figure, that of the observable Q or O)
Goal of the procedure	Calculating the resultant vector Finding the decomposition of the resultant into various directions	Finding information on the measurement of an observable (Q or O), by expanding a known state vector on the basis of its eigenstates
Constraints on the number of components and on their direction	No constraints: Superposition always has a physical meaning, independently of the number of forces and of their direction	The maximum number of components depends on the dimensions of the state space In usual situations, they are orthogonal to each other => the square (modulus) of each coefficient represents a transition probability
Interference	No interference	Always generating interference

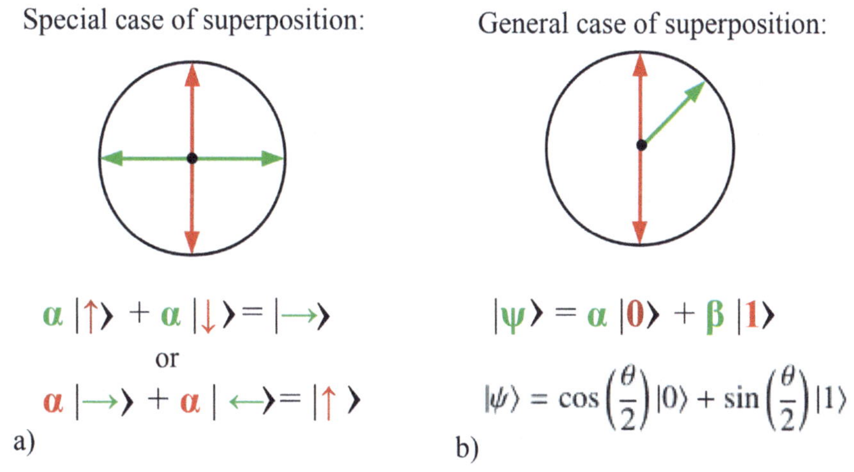

Fig. 2 Graphical representation of a two-state system in a Bloch circle combined **a** with a symbolic representation and **b** with the algebraic representation (Müller 2019). The correspondence of colors in the representations allows for making the appropriate connections and supports their correct interpretation

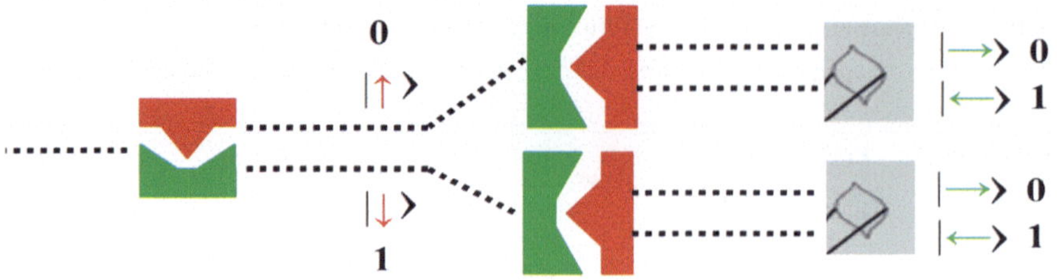

Fig. 3 Pictorial–symbolic representation of the thought experiment with two rotated Stern–Gerlach apparatus (Müller 2019). The dotted lines indicate the possible measuring results for the spin if measured according to the respective eigenstates of the corresponding Stern–Gerlach apparatus (Dür and Heusler 2016)

2.3.3 Uncertainty

The uncertainty often is explained by the thought experiment of three rotated Stern–Gerlach apparatus symbolizing the mathematical property of non-commutativity of spin operators. In Fig. 3, this is indicated by a combination of the pictorial, the symbolic and a numeric representation of this thought experiment.

The pictorial representation of the Stern–Gerlach apparatus and the sketches of the measurement results highlight the important aspect: The measurements after a corresponding rotation are with respect to different eigenstates. The results are then represented on the symbolic level with arrows and/ or Dirac notation. The numbers indicate that only certain values (0 or 1) can be attained in both measurements. This visualization describes a path from the physical interpretation of measurements to their mathematical formulation. A special important feature is that in both of the different bases (each related to one of the Stern–Gerlach apparatus) the possible results are coded by 0 and 1. To grasp this point is an important step in understanding on the one hand the physical appearance of uncertainty and on the other hand its mathematical structure and its formal description with help of the Dirac notation.

2.3.4 Probability

A central concept of QP is that a concrete measuring result can only be predicted with a certain probability. This physical fact can be represented graphically by a projection (left of Fig. 4). Its algebraic representation is described by the mathematical structure of the scalar product (squared) and is related to the symbolic representation with help of Dirac notation. Also, this example shows very clearly the interplay of mathematical elements and physical interpretation. In addition, it indicates that there are rules for the calculation or prediction of results and for their interpretation, including the syntax as well as the semantics of the Dirac notation.

Up to now, we had aspects related to the nucleus of QP (as described in part 3), indicating that a multiple representation is important for conceptual understanding.

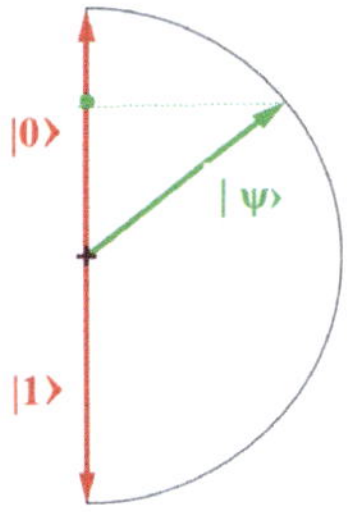

a) $P_{\uparrow}(\psi) = |\langle \uparrow | \psi \rangle|^2$

b) with $|\psi\rangle = \alpha\,|0\rangle + \beta\,|1\rangle$:

$$P_{|0\rangle}(\psi) = |\langle 0 | \psi \rangle|^2 = |\alpha|^2$$

Fig. 4 Different representations of the relation of measurement and probability (Müller 2019). On the left, a graphical representation of the measuring process as a projection is shown, on the right **a** the symbolic representation and **b** the algebraic one giving the value of the probability of the corresponding measuring result. The relation between them is supported by the colors

The following example is related to the "body" of QP (see part 3) and hence is important in anchoring the principles to a concrete example and conveying the message mentioned in the introduction of the importance of mathematics in QP.

2.3.5 Example

No-cloning theorem. The no-cloning theorem is important from two aspects: (a) practical: It is crucial for the security of quantum cryptography as it forbids the copying of quantum states. (b) didactic: It shows the limitation of purely symbolic or pictorial representations and qualitative descriptions, but the need of a rigorous mathematical treatment by algebraic representation and numerical calculation. In order to prove it, the physical situation has to be mapped onto a suitable mathematical structure, and at a certain point only after a calculation, the result is achieved.

First, a quantum copying machine Q has to be defined. This is done by translating the physical properties (e.g., Q as an unitary operator) with the help of a symbolic–algebraic representation: $Q|1\rangle|i\rangle = |1\rangle|1\rangle$ and $Q|0\rangle|i\rangle = |0\rangle|0\rangle$. Then, this definition together with the calculation rules is applied to an arbitrary state $|\psi\rangle = \alpha|0\rangle + \beta|1\rangle$. The calculation leads to a contradiction, hence showing that an arbitrary quantum state cannot be copied by a quantum copying machine. Herewith, the formalism takes on also a conceptual role in providing an insight that cannot be given without it (Wootters and Zurek 2009).

2.4 Implications from Linking of Representations

We have seen that the linking of different representations can enhance the learning of the concepts of quantum physics (Ainsworth 1999, 2008). At first, for each concept—superposition, uncertainty, probability, measuring process—suitable representations are chosen that support each other. For superposition, a graphical and a symbolic representations are linked, for the measuring process a pictorial and a symbolic one and for the probability a graphical, a symbolic and an algebraic representation. From here, it can be seen that the symbolic representation serves as a link between the different concepts and acts as a common anchor. In addition, the connection of several representations to the algebraic one opens the way to proceed step wise from concrete physics toward more abstract (algebraic) representations and hence enable the students to effectively work on quantum physical problems. In the TLP presented in the following section, we will see how the presentations offered in this section can be fruitfully used. Also teachers should be able to consciously use these different representations and be aware of their implications, their abstractness as well as of their advantages and disadvantages.

3 Preserving the Quantitative Nature of Quantum Physics Taught in High School[2]

Quantitative physics education has several important aspects. One of these is the potential it has to increase students' problem-solving understanding and competence in both disciplines (e.g., Galili 2018; Karam 2014; Kuo et al. 2013; Redish 2006; Tuminaro and Redish 2007) while increasing students' confidence if succeeded. Indeed, it is a double-edged sword, since lack of success in solving the problem might be very frustrating for the students; hence, a proper scaffolding of these skills is essential. Another aspect is that quantitative physics education is providing the students with evaluation means for feasibility of explanations or problems' solutions. It can also help students understand what is measured, and how the concepts were comprised. Especially in QP, mathematical sense-making can support students in developing a consistent interpretation of concepts and processes (see also Sects. 2 and 4). In addition, in light of deficiency of classical terminology for quantum phenomena, it may serve as a guide to understanding and represent an anchor and a resource for building mental models of the quantum world and even inspires physical concepts and insights (Sects. 2 and 4).

Yet, there is a great challenge in teaching QP quantitatively in high school, since usually the mathematics is too complicated. That is why many of those who teach QP in high school do it qualitatively or in a phenomenological manner. Whenever a quantitative approach is applied, it is mostly aimed at more advanced students, like 13th grade.

[2] This section is contributed by the PER group number 2.

3.1 Educational Context and Teaching–Learning Sequence

We planned a QP course for 12th grade students. The students' prior knowledge was algebra-based classical physics (Newtonian mechanics, electricity and magnetism and geometric optics). The course plan avoided complex numbers, as not all of our students were familiar with this concept. Due to lack of time (~30 h), the course was planned not following a historical path (Malgieri et al. 2017) but rather addressed a hierarchical structure of fundamental physical theories termed discipline-culture (Tseitlin and Galili 2005; Weissman et al. 2019). In this framework, every part of the theory is attributed either in the **nucleus**—the basic principles (like superposition, uncertainty and probability mentioned in Sect. 2), the **body** of the theory—implementation and examples (like the no-cloning theorem mentioned in Sect. 2) or in the **periphery**—alternative concepts that compete with the **nucleus**.

The first part of the curriculum and the experimental teaching included quantitative introduction to classical waves of light and matter, the photoelectric effect, the Bohr model of the atom, the Compton effect and De-Broglie wavelength. This might look like a historical path, but we taught these subjects very briefly, and mainly as the periphery of classical physics.

3.1.1 Conceptual Teaching Using Active Learning Methods

Striving for quantitative understanding, we did not want it to come at the expense of conceptual understanding. Therefore, the second part of the experimental teaching included an interactive teaching–learning sequence that immersed the students in the topics of the nucleus of QP. These topics were adapted for high school students through interviews with QP experts, physics education experts and several iterations of experimental teaching. We describe this part of the research elsewhere. These topics included the notion of *wavity* of matter next to spin, state, superposition, probability, wave function, measurement and the uncertainty principle. The notion of wavity designates the characteristic of a quantum physical object being spread among states (not as a "cloud of matter") either discrete or continuous. In the classical limit, wavity becomes a familiar distribution of matter in space. We used tasks that involved the students with text interpretation; questions implemented in short videos (like Reich 2019), experimentation with the PhET simulation of the Stern–Gerlach[3] experiment and class discussions.

3.1.2 Establishing Dirac Notation as a Formalism

The third part of the QP course took place after concepts were qualitatively established. In this part, we introduced the students to Dirac notation as a formalism that

[3] University of Colorado Boulder. PhET—Interactive Simulations. Available at: https://phet.colorado.edu/.

represents the concepts they had learned. This elaboration of the conceptual part is anchoring the nucleus principals in the way this theory is implemented, i.e., as part of the body of QP theory. We defined "state" using ket as a symbolic representation. We declared two kinds of states—"a pure state", which is a result of a measurement. It is denoted "$|a\rangle$" when a is the result of the measurement. The second state is a compound of basis states—a state of "superposition", which is denoted "$|\psi\rangle = A|a\rangle + B|b\rangle$". This relies on the conceptual understanding of the students that two pure states of the same physical entity (place, energy, etc.) are contradicting states, since in a measure we get just one value. We also presented the procedure of bra-ket squared ($(\langle\phi|\psi\rangle^2)$) as a technical procedure to calculate probabilities. We did not delve into the reason why this representation is a proper world description, and during later teaching, we justified it "experimentally" using the Stern–Gerlach PhET simulation quantitatively: We started with the simple situation of two states—the spin of electron. Rules were "extracted" from the simulation—given that spin was measured in the $X+$ axis the result was $+1/2$, another measurement in the same axis would give the same result in 100%, and vice versa, so$\langle x^+|x^+\rangle = 1$ and $\langle x^+|x^-\rangle = 0$ could be inferred (the negative result of the squared root of the bra-ket was excluded within the technical explanation). Once we had established that, the students found that the meaning of the coefficients is the probability to get the pure states, hence the "rule" is inferred. The students then "discovered" that spin in X-axis and Z-axis is under the principle of uncertainty—if the spin is a pure state in X-axis, then it is in superposition on the Z-axis. Since only two values of spin exist in a certain axis, the superposition expressions $|Z^+\rangle = \frac{1}{\sqrt{2}}|x^+\rangle + \frac{1}{\sqrt{2}}|x^-\rangle$ and $|Z^-\rangle = \frac{1}{\sqrt{2}}|x^+\rangle - \frac{1}{\sqrt{2}}|x^-\rangle$ could be established. Interestingly, it was quite easy for the students to construct in class the superposition expression for $Z+$. However, when they were asked to construct the expression for $Z-$, they also constructed the same expression as $Z+$. We then asked them to calculate $\langle z^+|z^-\rangle$, as they wrote them, knowing that it should yield 0, like any contradicting pure states, yet the students found it did not. We asked the students what could be done, so the mathematics representations will describe the physical world, i.e., would yield $\langle z^+|z^-\rangle = 0$ and they came up with the idea of changing the sign of one of the components in the Z-expression. This way of teaching sustains the calculable nature without stumbling over too complicated mathematics that does not fit the level of the students. Further on, we used the simulation to construct the representation to get spin $+$ in any angle from the X-axis. The students got the graph in Fig. 5 from the simulation, and they needed very little help to come up with the understanding of the cosine squared, and finding half the angle (Fig. 6). At that point, the students derived the expressions $|\theta^+\rangle = \cos\frac{\theta}{2}|x^+\rangle + \sin\frac{\theta}{2}|x^-\rangle$ and $|\theta^-\rangle = \sin\frac{\theta}{2}|x^+\rangle - \cos\frac{\theta}{2}|x^-\rangle$ from the simulation. Another activity that we did with the students in order to consolidate the body of QP in other implementations is demonstrating polarization (Malus's law) and framing it in the terminology of superposition of states, including the formulas (finding that in this case it is θ and not $\frac{\theta}{2}$). Another implementation was BB84 protocol for quantum cryptography that was introduced as the outcome of the two-state understanding of light polarization. This third part took about 8 h.

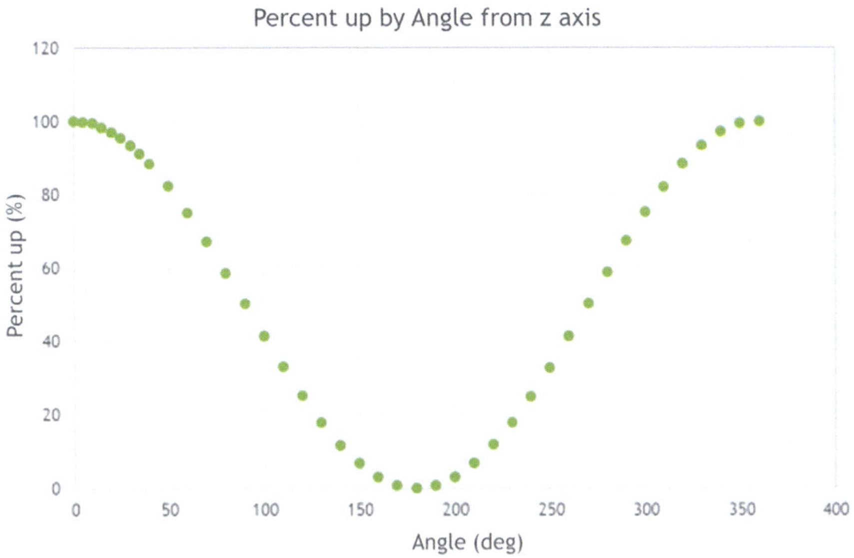

Fig. 5 Results of measures in Stern–Gerlach PhET simulation

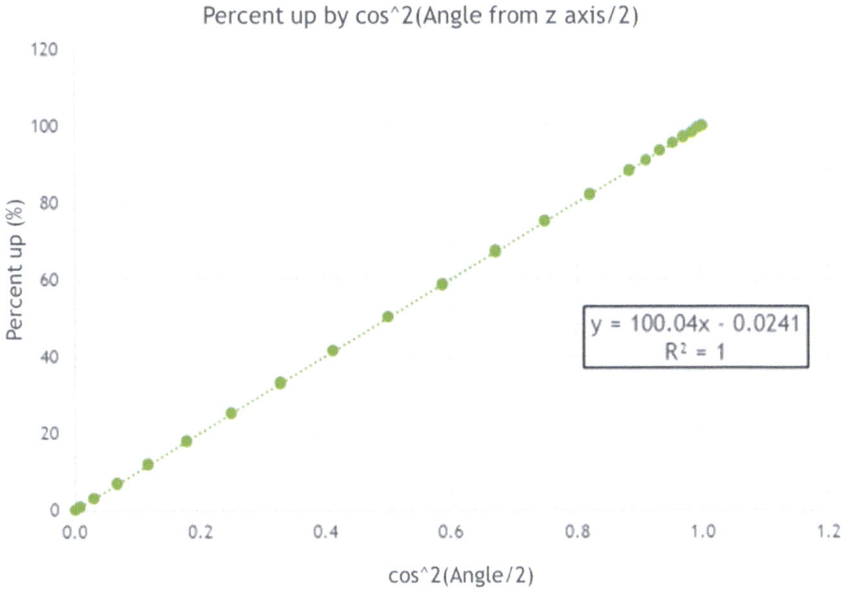

Fig. 6 Students' analysis to reach the dependency of the angle

3.2 Research Questions

- Can teaching based on the Dirac notation approach be effective in leading students to understand basic concepts like “state”, “superposition”, “wave-particle duality”, “wave function” and the probabilistic nature of QP?
- Can teaching based on the Dirac notation approach be helpful in anchoring the concepts in quantitative problem solving?

3.3 Method

The teaching–learning sequence (~8 h) included frontal explanation in class followed by whole class training. Then, two work-alone exercises were administered, and a final test was implemented. The goal of the first exercise was to make sure the students understood the technical procedure and the meaning of its result. Hence, they were asked to calculate the expected results of certain measurements made in the Stern–Gerlach experiment and compare it to the simulation. As offered above (Sect. 2), this part used a symbolic form (see Figs. 1d, 2a above) and was a lever for the algebraic form (Fig. 1f) in the second exercise. The second exercise was administered after learning about classical light polarization. Using Google forms, the students were asked a question (example below) and could choose to submit a response or to get a clue. Choosing the latter allowed them another similar choice (for maximum of four clues). Three kinds of responses were required: (a) uploading a full calculation of the student’s answer, (b) verbal explanation of the meaning of the results and (c) numerical final solutions. Feedback, after submission, was a link to a full solution.

The test included questions aimed to measure both conceptual understanding of the use of the notation and the practical ability to implement it quantitatively (see below). We analyzed students’ responses to the test, and some students were interviewed. The lessons were video recorded for further analysis.

3.4 Results

3.4.1 Dirac Notation as a Tool for Problem Solving

In order to test for the conceptual understanding while preserving the quantitative nature of problem solving, we asked 12th grade students who participated in the course the following question in the final test:

A state of an electron is given by $|\psi\rangle = 0.6|x^+\rangle - 0.8|x^-\rangle$. The state $|\varphi\rangle$ is a state in which there is 0.81 probability to get positive spin in the X-axis.

1. Write the state $|\varphi\rangle$ as a superposition of the eigenstates of spin x. If there is more than one option, write them all.

2. It was measured and found that the state of the electron is $|\psi\rangle$. What was the measure and what was the result of it?
3. It was measured and found that the state of the electron is $|\psi\rangle$. After this measure, what is the probability that if we measure we will get $|\varphi\rangle$?

In part 1, the students were asked to present: (i) Skill of using the notation for superposition in a certain base ($x^{\pm}$), (ii) Understanding of the meaning of the coefficients and the probabilities and (iii) Understanding that there is more than one superposition state with the same probabilities. Out of 26 students, 23 (88.46%) used the notation correctly, even if some students made some calculation mistakes for the coefficients. Eighteen students (69.23%) answered this section correctly, though four of them did not write another superposition state.

In part 2, the students were asked to present that they understand the meaning of a measure though it is presented as a state of superposition and connect it to an angle. Out of the 26 students, 21 (80.77%) could tell that the measure was of spin and 16 students (61.54%) could tell the sign. Fourteen of the students (53.85%) could connect the measure to an angle from the X-axis, but only nine students (34.61%) gave a full answer to this section and calculated the angle correctly. It seems that one difficulty that students might have is the transformation between bases. Apparently, this should be addressed explicitly in teaching.

In part 3, the students were asked to present the skill of using the notation to calculate probability. Most of the students, 22 out of 26 (84.61%) followed the *bra-ket squared* procedure and made a proper response.

3.4.2 Dirac Notation as a Conceptual Representation

In order to see what the students infer from the use of Dirac notation as a representation, we asked the students in the final test the following open question:

Does the expression $|Z^{+}\rangle = \frac{1}{2}|a\rangle + \frac{\sqrt{3}}{2}|b\rangle$ represent a wave? Explain.

A group of 17 students of 12th grade were asked this question. All said it does represent a wave. Students used mainly two concepts to explain why this expression is a wave: All but one student (94.11%) referred to the representation of *superposition.*

An illustrative answer was:

Eilon: *This expression represents a wave because this is an equation that describes* **superposition** *the superposition is based on the* **wavity of matter**, *and therefore, this expression represents a* **wave behavior** *or a wave.*

This response implies that students shifted the meaning of "wave" to "superposition" and adopted the new meaning in QP.

Eleven students (64.7%) referred to the *probability* to get different results in different measures.

An illustrative answer was:

Yotam: *This expression represents a wave indeed. It can be seen because it describes* **some eigenstates** *of a physical quantity/certain physical property. I mean,* *this represents a state of* **superposition, "a map of probabilities"** *of a certain physical quantity, and this is why it represents a* **wave function** *of a particle (wave function is a map of probability of a physical quantity).*

From this and former examples, it can be inferred that the student grasped the main meaning of quantum wavity as a superposition of states. In addition, it seems that they can interpret the formalism and extract the properties of a measurement. Reading Yotam's answer carefully, one might even consider near transfer from a discreet world of spin and slits to a continuum, as he describes the wave function as "a map of probability" to get a measurable value of a physical quantity, but this would require further examination.

4 Interplay Between Math and Physics in an Educational Path on Quantum Mechanics in Secondary School[4]

4.1 Research Perspectives for the Consolidation of Quantum Concepts Through Mathematics

Addressing the challenge of teaching quantum concepts that have an in-built mathematical component requires two main tasks: developing an insight into the nature of student difficulties with math in QP and developing suitable learning strategies for the consolidation of concepts through math.

As concerns the first task, given the lack of data on secondary school students, we analyzed the comprehensive review of Singh and Marshman (2015) on difficulties at upper-undergraduate level. We are aware that advanced university students differ in many aspects from the population of our interest. Nevertheless, we rely on the results found by the authors, who make a case for strong analogies between patterns of difficulties in introductory physics and upper-division QP.

The lens we used to examine the review is the framework on the role of math in physics developed by Uhden et al. (2012). This framework posits a deeply tangled unity of mathematical and physical models and identifies two roles played by math in physics: one technical and the other structural, the latter referring to the role of math in structuring physical entities and situations that emerge in the processes of interpretation and formalization. By reading the review through the lens of this framework, we observe that deep and widespread mathematical difficulties elicited by research are strongly linked to the structural role of mathematics within the theory (Uhden et al. 2012). In other words, the primary problem with math in QP is not its technical sophistication, but the representational shift affecting formal entities already familiar to students, such as vectors and their superposition, in passing from a classical to a quantum perspective, and the adoption of new formal entities to represent basic but non-intuitive notions.

[4] This section is contributed by the PER group number 3.

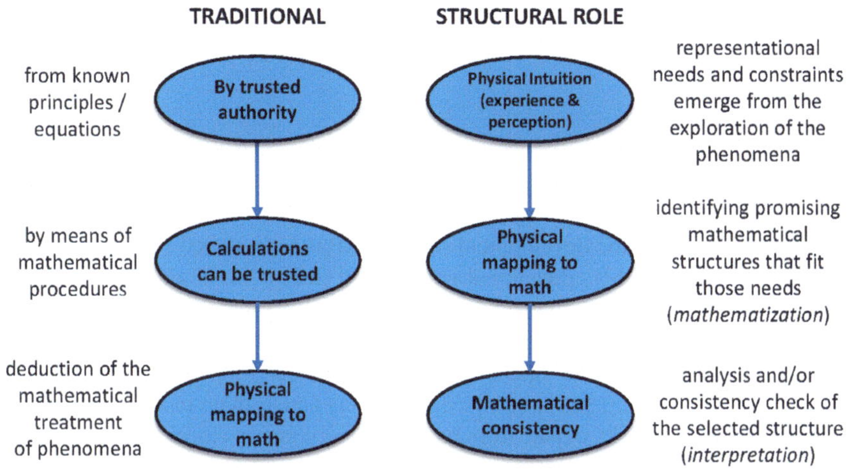

Fig. 7 Comparison between a traditional instructional sequence and one highlighting the structural role of mathematics, based on the chain of activation presented in Redish and Kuo (2015)

This fact alone provides a first hint on how to address difficulties concerning known entities that supply structure to quantum concepts: Instructors and researchers need to analyze them in advance from a structural standpoint, focusing on the representational-shifts affecting them, as they may reasonably involve challenges to students.

However, as the structural role of math in physics emerges in the processes of mathematization and interpretation, it is necessary to develop strategies in order to operate an effective transition, back and forth, between observations/predictions in the lab and the formal representation of the relevant concepts. As concerns the identification of such strategies, we analyzed studies on mathematical sense-making in physics (e.g., Uhden et al. 2012; Redish and Kuo 2015).

In Fig. 7, we illustrate a comparison between a traditional instructional sequence on the interplay math–physics and one highlighting the structural role of math. The latter goes from physical issues to mathematics (from concrete to abstract) followed by a new interpretive activity, aimed at clarifying further physical implications of the newly introduced structure. In the case of QP, the representational shift in formal entities already familiar to students adds to the importance of the interpretive phase. The strategies used in our TLP are variants of the structural sequence described.

4.2 Implementation in the Context of the Linear Polarization of the Photon and the Hydrogen Atom

In order to illustrate how we used the aforementioned guidelines in a classroom environment, we describe some general feature of the TLP and its refinement process by means of design-based research (DBR) cycles (Barab and Squire 2004), focusing on design and revision in the introduction of the quantum state and its formal representation.

4.2.1 Overview of the TLP

The starting point for the development of this TLP is represented by the Ghirardi–Grassi–Michelini (GGM) educational path on QP (Michelini et al. 2004; Ghirardi et al. 1995). GGM shares with the new proposal an inquiry-based approach (Pedaste et al. 2015), the context of linear polarization, the use of cheap experimental tools such as polarizing filter and calcite crystals and the adoption of worksheets as a guide for supporting students in knowledge building and as main instrument for data collection. The two TLPs depart from each other as concerns their focus (state and superposition in GGM, the relations between properties in the new TLP (Zuccarini 2018), i.e., the rules that determine the acquisition, the loss and the retention of definite values of observables in the measurement process), educational strategies (e.g., the new TLP largely adopts strategies for mathematical sense-making), the sequence (GGM proceeds from the state to measurement, and the new TLP goes the other way around) and part of the context (in the new TLP, each time that new content is introduced, we discuss first the simple context of polarization, but then move on to examine the theoretically significant case of the hydrogen atom).

4.2.2 Refinement of the TLP

In the years 2014–2018, the TLP was tested on high secondary school students attending the Udine Summer School of Modern Physics, and was subsequently revised. The sample consisted of good performing students from different Italian regions, whose number varied each year from a minimum of 29 to a maximum of 41. In 2019, the TLP was tested on 18 motivated students from Liceo Galilei, Trieste, taking optional physics classes.

The instructional sequence on QP requires only a knowledge of introductory Newtonian mechanics and of basic vector algebra (superposition and scalar product). All of the participant students were enrolled either in the last or the second to last year of secondary school and had already covered the prerequisite topics at school in previous years.

Data sources consisted of written answers to worksheet questions, occasionally enriched by notes taken by researchers during teaching experiments. Data were analyzed by qualitative research methods (e.g., Erickson 2012). A-priori categories were built by identification of crucial conceptual contents and literature analysis on difficulties in QP. The categories were revised on the basis of the elements introduced by student answers. The analysis aimed to identify emerging element clusters in student reasoning.

4.3 *Introduction of the Quantum State and of Its Formal Representation*

The notion of state is an emblematic example of a known concept that undergoes a shift in meaning with the transition to the quantum picture. The classical state identifies the condition of a system at a given instant, i.e., the set of values of physical quantities describing the system. The quantum state identifies the behavior of the system in measurement at a given instant, i.e., the set of probability distributions for measurement outcomes of each observable. The quantum state is represented by a vector, an algebraic entity, secondary students are familiar with. However, if we analyze this entity from a structural perspective, we immediately see how misleading this statement can be. Actually, students are only acquainted with vectors representing physical quantities and lying in the lab space. The state vector, instead, is an abstract unit vector, defined up to a phase, and lying in a Hilbert space. In addition, the features of quantum superposition are quite different from those of a superposition of vector physical quantities, e.g., forces (Table 1).

If we address the concept of quantum state starting from the vector, students may easily interpret it as the structure they know. In the context of linear polarization, this interpretation is even more enticing, as there is a relation preserving isomorphism between the property space and the state space (Fig. 8).

Here, we describe the instructional sequence developed to introduce the quantum state and the basic properties of the state vector (modulus, direction, dimensions), limiting ourselves—for reasons of space—to the part related to the linear polarization of the photon. This context allows us to focus on the elementary aspects of the stochastic nature of quantum measurement, i.e., transition probabilities, since contrarily to most measurements, the concept of expectation value is not particularly significant. The sections concerning the state of the hydrogen atom and the features of quantum superposition will be presented in future works.

In order to promote a consistent understanding of the concept and the formal representation of the state, we designed this sequence: (1) concept + physical information encoded in the state, (2) vector mathematization, (3) vector interpretation.

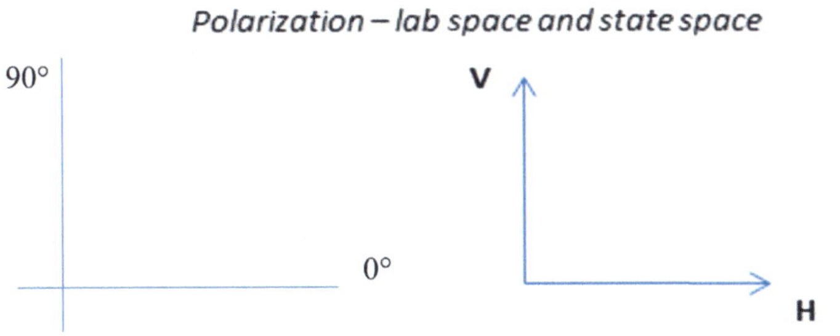

Fig. 8 Relation preserving isomorphism between the property space and the state space: The angles between polarization properties are equal to the angles between the corresponding states

After addressing step (1) by means of a short discussion, highlighting the bijection between linear polarization states and properties, we move on to step (2).

Students already know the interpretation of Malus's law in terms of single photons, i.e., the probability that a photon prepared with a property at θ passes a filter with axis at φ: $p(\theta \rightarrow \varphi) = cos^2(\theta - \varphi)$, covered while discussing measurement. Now, as the quantum state describes the behavior of the system in measurement, the instructor states that our goal is to describe the transition rule from one state to another by means of simple operations (sums/products) between the entities representing each state, thus fostering an algebraic form of reasoning instead of a visual analogy. We ask:

Item 1 (oral). How many kinds of algebraic entities do you know? Make a proposal and explain why it could represent the state of polarization of a photon.

In Udine Summer School 2018, students answered this question in a whole class discussion. They mentioned numbers, vectors and matrices. Since angles are the only significant aspect discussed in polarization, but numbers lead to a transcendental function, they propose vectors "in the same direction as the property". After having identified a suitable candidate for the formal representation, we ask:

Item 2 (worksheet). Propose an operation between a generic state vector and a state vector that gives a consistent result concerning the transition probability (Fig. 9).

Step (3) is implemented first by means of interpretive questions aimed to facilitate student's identification of the representational and conventional elements of the state vector. When students recognize that we cannot ascribe a meaning to vectors of different lengths, the transition rule is reduced to $(U \cdot W)^2$. As concerns opposite directions, they establish that U and -U represent the same state either by using a physical reasoning on polarization ("as the property is identified by a direction") or a global math reasoning ("as the result is a square of a dot product").

The final question is the following:

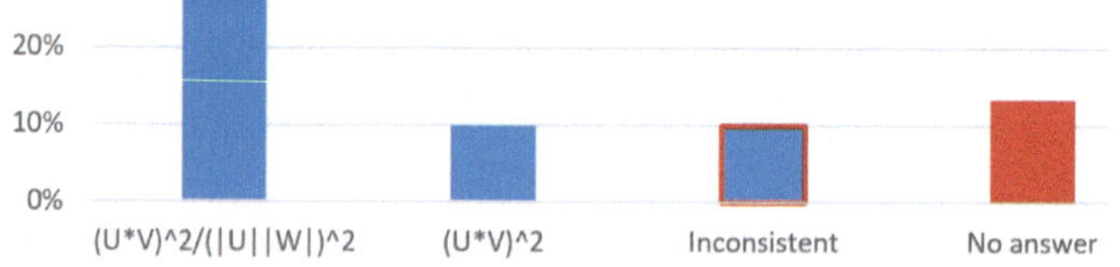

Fig. 9 Results of item 2 in Udine Summer School 2018: On 30 students, 23 made consistent proposals concerning the operation at hand. As the length of the state vector has not been addressed yet, most of them came up with the formula: $(U \cdot W)^2/|U|^2|W|^2$

Item 3 (worksheet). A physical quantity is a quantitative feature of a system or process whose value can be determined by measurement. Is the quantum state a physical quantity?

The results are displayed in Fig. 10. Half of the students (15/30) consistently answered no, and the state (vector) is abstract as it represents "a set of probability distributions" or "a probability". However, despite the careful design of the sequence, 8/32 students identified the quantum state as a physical quantity, mostly by activating the conceptual resource that the state concerns measurement, with relation to the nature of the physical concept ("it is measurable", "it represents probability in measurement"). This shows how tricky is the context of polarization as concerns interpreting the nature of the state.

As a consequence, in order to guide students to productive reasoning, we revised the instructional sequence by inserting, before item 3, a further interpretive question concerning the physical dimensions of the vector ("determine whether the state vector has physical dimensions and eventually which ones"). Immediately after the new item, we added a discussion of the state of a hydrogen atom and the simplified treatment of its state vector (real vector with the same structural features as the polarization state vector). The transition to the new context removes the ambiguity in the relation between state and property, since four compatible properties belonging, e.g., to the observables must be used to identify the state. A more detailed description of this part of the TLP and related items will be presented in future works.

The new sequence was implemented in 2019 on 18 volunteer students from Liceo Galilei, Trieste. The worksheet item on the dimensions of the state vector was correctly answered by all students, most of them (11) stating that it has no dimensions "because it gives a probability, a pure number". This time, also item 3 was largely successful, as shown in Fig. 11.

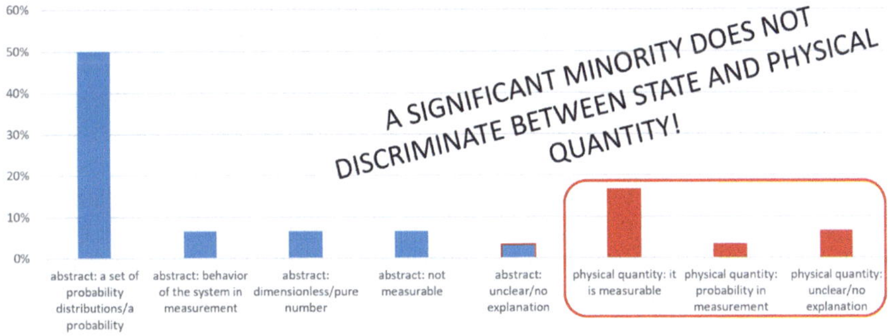

Fig. 10 Results of item 3 in Summer School 2018

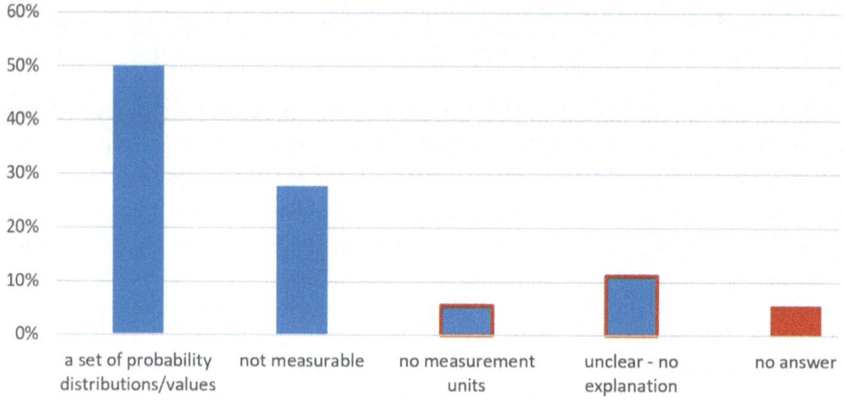

Fig. 11 Results of item 3 in Liceo Galilei, 2019. All students but one recognized the abstract nature of the state, most of them with consistent reasoning

5 Conclusion

We have illustrated how, through the adoption of different representations and the design of carefully structured connections between them, the basic mathematical formalism of QP can be put to the service of conceptual construction in a high school context. In particular, we have shown how the symbolic representation serves as a link between the different concepts and how the connection of several representations to the algebraic one paves the way to the possibility of making predictions.

The data of both TLP show that instructional sequences based on an effective transition between the concepts involved and their formal representations in terms of mathematizing and interpretive activities performed by the students can be largely successful in helping them to appreciate the structural aspects of the mathematical entities: their representational and conventional features, as well as the way in which physical information is encoded in those entities. In order to significantly improve conceptual building based on math, there is a need of a preliminary analysis of the structural challenges hidden in each mathematical entity (e.g., the issues concerning the polarization state vector) and refinement cycles of the TLP based on data.

Especially the first TLP, described in Sect. 3, implies that the use of Dirac notation together with conceptual teaching is helpful in teaching QP in high school: It supports students to formalize the concepts of "state", "superposition", "wave-particle duality", "wave function" and the probabilistic nature of QP. The symbolic math allows quantitative problem solving; whereas other approaches require much more complex mathematics. Moreover, even though students might consider the Dirac notation as a technical means to express the conceptual ideas of QP, the symbolic math seems to anchor the concepts in practice and assist students in understanding the experimental way QP is justified. Some implementations of QP like BB84 protocol of quantum encryption, quantum coin toss, the Stern–Gerlach experiment and the Mach–Zehnder experiment are especially appropriate for quantitative examination of the concepts usually taught qualitatively. The developed representations and their interlinking could be used in a quantum cryptography course for high school students, to evaluate whether and how their adoption as thinking tools facilitates conceptual understanding, the production of qualitative argumentation and of quantitative prediction. It might be that Dirac notation is a useful basis for the transfer from discrete to continuum systems, yet there is a need of further research and development of TLPs in order to evaluate its effectivity. In this, the visualization by combining different representations might be a fruitful means.

References

Ainsworth S (1999) The functions of multiple representations. Comput Educ 33:131–152

Ainsworth S (2008) The educational value of multiple-representations when learning complex scientific concepts. In: Gilbert JK, Reiner M, Nakhleh M (eds) Visualization: theory and practice in science education. Springer, Dordrecht, pp 191–208

Airey J, Linder C (2009) A disciplinary discourse perspective on university science learning: achieving fluency in a critical constellation of modes. J Res Sci Teach: off J Natl Assoc Res Sci Teach 46(1):27–49

Barab S, Squire K (2004) Design-based research: putting a stake in the ground. J Learn Sci 13(1):1–14

Dür W, Heusler S (2016) The qubit as key to quantum physics part II: physical realizations and applications. Phys Teach 54(3):156–159

Erickson F (2012) Qualitative research methods for science education. In: Second international handbook of science education. Springer, Dordrecht, pp 1451–1469

Galili I (2018) Physics and mathematics as interwoven disciplines in science education. Sci Educ 27(1–2):7–37

Galili I (2019) Towards a refined depiction of nature of science. Sci Educ 28(3–5):1–35

Geyer MA, Kuske-Janßen W (2019) Mathematical representations in physics lessons. In: Pospiech G, Michelini M, Eylon BS (eds) Mathematics in physics education. Springer, Dordrecht, pp 75–102

Ghirardi GC, Grassi R, Michelini M (1995) A fundamental concept in quantum theory. In: Thinking physics for teaching. Springer, Boston, pp 329–334

Karam R (2014) Framing the structural role of mathematics in physics lectures: a case study on electromagnetism. Phys Rev Spec Top-Phys Educ Res 10(1):010119

Krijtenburg-Lewerissa K, Pol HJ, Brinkman A, Van Joolingen WR (2017) Insights into teaching quantum mechanics in secondary and lower undergraduate education. Phys Rev Phys Educ Res 13(1):010109 1–21

Kuo E, Hull MM, Gupta A, Elby A (2013) How students blend conceptual and formal mathematical reasoning in solving physics problems. Sci Educ 97(1):32–57

Lemke JL (1998) Reading science. In: Martin JR, Veel R (eds) Critical and functional perspectives on discourse of science. Routledge, London, pp 87–113

Malgieri M, Onorato P, De Ambrosis A (2017) Test on the effectiveness of the sum over paths approach in favoring the construction of an integrated knowledge of quantum physics in high school. Phys Rev Phys Educ Res 13(1):010101

Michelini M, Ragazzon R, Santi L, Stefanel A (2004) Discussion of a didactic proposal on quantum mechanics with secondary school students. Il Nuovo Cimento C 27(5):555–567

Müller Ch (2019) Planung einer unterrichtseinheit zur quantenkryptografie und akzeptanzanalyse der dirac-notation. Master thesis. TU Dresden

Pedaste M, Mäeots M, Siiman LA, De Jong T, Van Riesen SA, Kamp ET, Manoli CC, Zacharia ZC, Tsourlidaki E (2015) Phases of inquiry-based learning: definitions and the inquiry cycle. Educ Res Rev 14:47–61
Pospiech G (1999) Teaching the EPR paradox at high school? Phys Educ 34(5):311
Pospiech G (2003) Philosophy and quantum mechanics in science teaching. Sci Educ 12(5–6):559–571
Redish EF, Kuo E (2015) Language of physics, language of math: disciplinary culture and dynamic epistemology. Sci Educ 24:561–590
Redish EF (2006) Problem solving and the use of math in physics courses. arXiv preprint physics/0608268
Reich H (2019) https://www.youtube.com/watch?time_continue=7&v=Q_h4IoPJXZw. Last assessed 2019/1/24
Singh C, Marshman E (2015) Review of student difficulties in upper-level quantum mechanics. Phys Rev Spec Top-Phys Educ Res 11(2):020117 1–24
Tseitlin M, Galili I (2005) Teaching physics in looking for itself: from a physics-discipline to a physics culture. Sci Educ 14(3–5):235–261
Tuminaro J, Redish EF (2007) Elements of a cognitive model of physics problem solving: epistemic games. Phys Rev Spec Top-Phys Educ Res 3(2):020101
Uhden O, Karam R, Pietrocola M, Pospiech G (2012) Modelling mathematical reasoning in physics education. Sci Educ 21:485–506
Weissman E, Merzel A, Katz N, Galili I (2019) Teaching quantum mechanics in high-school—discipline-culture approach. J Phys: Conf Ser 1287(1):012003. IOP Publishing
Wootters WK, Zurek WH (2009) The no-cloning theorem. Phys Today 62:76–77
Zuccarini G (2018) Ph.D. thesis, University of Udine

Introduction of Contemporary Physics to Pre-university Education

Jerneja Pavlin, Alberto Stefanel, Philippe Lindenau, Michael Kobel, Anja Kranjc Horvat, Jeff Wiener, Sascha Schmeling, Andreas Borowski, Dagmara Sokołowska, and Mojca Čepič

Abstract This symposium discusses the modes and means for introduction of contemporary physics to pre-university level education illustrated by a few successful examples. Examples come from soft matter physics in liquid crystals and hydrogels and solid-state physics in superconductors. The role of analogies is discussed and illustrated in analogy between superconductors and particle physics. In addition, a development of a professional development programme that will bridge the gap to new knowledge for active in-service teachers is presented.

Keywords Contemporary physics · Pre-university education · Physics education

1 Introduction

Pre-university physics is often considered as old and boring (Osborne et al. 2003a). However, scientists in laboratories and behind computers occupied with discoveries of new phenomena have a diametrically opposite opinion. For them, physics is a

J. Pavlin · M. Čepič (✉)
Faculty of Education, University of Ljubljana, Kardeljeva pl. 16, Ljubljana, Slovenia
e-mail: mojca.cepic@pef.uni-lj.si

A. Stefanel
Physics Education Research Unit, University of Udine, Via delle Scienze 206, Udine, Italy

P. Lindenau · M. Kobel
Institute for Nuclear and Particle Physics, Technische Universität Dresden, Dresden, Germany

P. Lindenau
Didactics of Physics, Technische Universität Dresden, Dresden, Germany

A. K. Horvat · J. Wiener · S. Schmeling
European Organisation for Particle Physics, CERN, Genève 23, Switzerland

A. K. Horvat · A. Borowski
University of Potsdam, Karl-Liebknecht Str. 24/25, Potsdam, Germany

D. Sokołowska
Smoluchowski Institute of Physics, Jagiellonian University, Lojasiewicza 11, Krakow, Poland

B. Jarosievitz and C. Sükösd (eds.), *Teaching-Learning Contemporary Physics*, Challenges in Physics Education,
https://doi.org/10.1007/978-3-030-78720-2_5

vivid and relevant science with interesting challenges, full of surprises and beauty. Why such a difference in opinion? Several reasons were already discussed, from a general public opinion that physics is very difficult and only very smart, however rather strange people, can understand it, to the modes of teaching often focused on abstract calculation of problems irrelevant for everyday life, and rigid curricula that consider only topics more than hundred years old, an eternity from the standpoint of students. In spite, students are aware that science is important for the development of technology and that technology nowadays is based on relatively recent findings.

What do we call "contemporary physics" in this contribution? Contemporary physics is used for research topics that are actively investigated, have regular and frequent contributions in scientific journals, and conferences focused on these topics. Besides, it is welcome if the context is provided by devices used every day that are connected to new findings in those fields or other means like science fiction or sustainability problems.

Introduction of the front-end physics to the physics classroom at the pre-university level meets several obstacles and is a difficult nut to crack (Čepič 2017). Why is that so? New findings in physics are recent, which means an average teacher was not informed on such topics during the pre-service education, as the knowledge has not existed yet. Next, researchers at the front end usually do not have enough time, lack ambition, but they also do not have the means and the knowledge to adapt the advanced science to the cognitive level of pre-university students. Finally, educators are usually not in everyday contact with the contemporary physics and are occupied with problems relevant for teaching and learning the topics in existing curricula. Nevertheless, we believe that the front-end science can be and has to be introduced to students at the pre-university level, and its relevance for technology we meet every day has to be demonstrated. To successfully introduce the front-end science to the physics classroom, researchers, educators and in-service teachers have to collaborate. In this contribution, we present a few examples of successful approaches to the introduction of contemporary physics, which demonstrate that it is worth the effort.

First, we focus on two examples from soft matter physics. A material that is closely related to our everyday life is liquid crystals, which are a crucial part of every liquid crystal display. Such displays are found everywhere, from mobile phones and pads to regular computer and TV screens. Everybody is familiar with diapers that keep the skin dry even when toddlers sit on them. The responsible materials are hydrogels that are effective absorbent of water but they do not release the water easily. The physics of these new materials was introduced to the first year of pre-service primary school teachers, the population that has just begun the study and could be considered very similar to high school students in their final year.

From soft matter, we continue to solid-state physics and introduction of superconductivity to the high school. We demonstrate its effects to several hundreds of students by the study that included several high schools.

As it is well known, analogies are an important and useful tool in physics. Especially when analogies link two apparently very different subfields in physics, their importance becomes even more evident. An analogy that links the short-ranged weak interaction of the Standard Model of particle physics with electromagnetic superconductivity in solid-state physics is presented and its usefulness in the context of a teaching approach that focuses on charges as the major concept of particle physics is discussed.

However, one should not forget that the education begins and ends with enthusiastic, well informed and confident teachers. The contemporary physics is obviously out of teachers' comfort zone. To provide the teachers with the new knowledge, and to increase her/his confidence regarding contemporary physics, additional in-service trainings are needed. The last part discusses the development of professional development programmes at CERN that will provide the training and the consequent confidence in contemporary topics to new generations of in-service teachers.

2 Liquid Crystals and Hydrogels as Examples of Soft Matter

Physics should present the bridge between innovations and applications. Soft matter is a relatively new research field, where polymers, gels, liquid crystals, but also cells, and other biomaterials are investigated. Although a wide variety of materials qualify as soft matter, for introduction of contemporary physics to the classroom we focus on two examples, hydrogels and liquid crystals.

Botanist Friedrich Richard Reinitzer was in 1888 experimenting with cholesteryl benzoate and discovered properties of liquid crystals. However, it was Otto Lehmann, who addressed the question of the phase of matter which led to the science of "liquid crystals". Liquid crystals are a state of matter that has properties characteristic for conventional liquids and those of solid crystals. Liquid crystals are in the liquid crystalline phase in several products, from welding helmets to cosmetic products. However, the most common product with liquid crystal is a liquid crystal display (LCD), an essential part of laptops, iPods, mobile phones, etc. (Dunmur and Sluckin 2011).

Another recent material is hydrogels, which consist of a network of polymer chains that are hydrophilic. They are used in products that improve everyday life (e.g. diapers, watering beads for plants, perfume delivery, cosmetics, tissue engineering, etc.) due to the ability of loading, retaining and releasing fluids (Okay 2009; Calo and Khutoryanskiy 2015). They differ from sponges because of resistance to stress and water retention under the pressure that is important for diapers for example.

How contemporary the science (physics) of liquid crystals and hydrogels is might be shown by the number of publications. If one uses the keyword "liquid crystal*" in Web of Science, the result is up to 10 hits per year till 1963. Afterwards, the number of publications increases and in 1969 more than 100 papers were published to more than 1000 papers in 1989. In 2019 (till 27th October 2019) 3026 papers were published. Similar results are obtained with a keyword "hydrogel*". Up to 10 hits per year until 1972, when number of publications per year has started to increase. In

1990 more than 100 papers were published, in 2003 more than 1000 papers, and in 2019 (till 27th October 2019) 8323 papers, which indicates that the research in both fields is vivid.

Both topics are therefore good candidates for introduction of contemporary physics to the classroom. To prepare the teaching modules we considered a preliminary knowledge of students, detailed learning objectives, and developed simplified hands-on experiments (Pavlin and Čepič 2017).

The study of existing knowledge about liquid crystals and hydrogels of the 1st-year pre-service primary school teachers indicated that it is very limited or mostly absent. Some students were aware of the term and were able to list a few products or their properties. On average, they achieved 16% of possible points at liquid crystals prior knowledge test and 7% on hydrogels knowledge test (Pavlin and Čepič 2017; Pavlin et al. 2013).

Having these results in mind we set three aims of the study:

- How to design the teaching module on liquid crystals and hydrogels?
- What knowledge did the pre-service teachers gain after the module?
- What kind of a feedback did pre-service teachers give on the modules on liquid crystals and hydrogels?

2.1 Modules

The module on liquid crystals was implemented in the 1st year of pre-service elementary teachers' programme. The students enrol to this programme after finishing the upper secondary school with a general programme called "Gimnazija" at the age 19–20. This programme was chosen because the students attending it have a wide variety of interests and are usually not very motivated for science, which makes the population similar to the high school students. The module on liquid crystals had three parts: a lecture, a chemistry laboratory, and a physics laboratory, for duration of 90 min each. During the lectures, students got familiar with basic properties of liquid crystals, optical properties, and LCD's technology as LCD was emphasized as a science context. Synthesis of a liquid crystal called MBBA is carried during laboratories in chemistry, and subsequently at laboratories in physics, students performed hands-on experiments with it. More detailed, they studied liquid crystalline phase as an additional thermodynamically stable phase, and the phenomena related to liquid crystals as double refraction and colours. During the activities, students return to the LCD and explanation of its functioning.

Due to the similarity of the 1st year students of pre-service elementary teachers' programme with high school students, also the module about hydrogels was implemented in this group, but not in the same school year, due to the time limits imposed from the science course. The teaching module consists of two parts, a lecture and a physics laboratory, for duration of 90 min each. Basic properties of hydrogels with emphasis on the mechanism of swelling, optical properties, and mechanical properties were presented during the lecture. During physics laboratory, students performed hands-on experiments using hydrogel pearls. They studied to more detail: absorption of water, density of hydrogel pearls, optical properties of hydrogels (refractive index, refraction, lenses, fluorescence, etc.), the hydrogel pearl in different media, and hydrogels under the pressure.

2.2 *Implementation Studies and Results*

The sample of evaluation study, that is, the 1st-year pre-service primary school teachers were on average 20 years old ($SD = 0.5$), mostly female students (96%) and with no specific interest in science. Data was collected by knowledge test (pre-post design) and a questionnaire including general data and open-ended questions about the opinion on the LCs/hydrogels teaching modules during the obligatory course Science, Physics (liquid crystals in Chemistry as well) contents (lectures, laboratory work). All data were collected and statistically processed in Excel and a qualitative procession of data at open-ended questions was done. Basic descriptive statistics was used.

Tables 1 and 2 present students' achievements on the post-test after the implementation of the modules. It is evident that students increased the non-existing prior knowledge about liquid crystals and hydrogels.

The majority of students (75%) gave positive feedback to the teaching modules about liquid crystals and hydrogels. For insight into views of students, a few students' responses on the teaching modules about liquid crystals and hydrogels are listed.

Liquid crystals:

- *Physics has meaning, as shown.*
- *The best thing was the work with a microscope.*
- *The experiments were interesting.*
- *It is good to have some general knowledge about how stuff works. I wish we could learn about it more often.*

Table 1 Students' achievements on the post-test immediately after the implementation of the teaching module on liquid crystals

Question about LCs	Percentage of students with correct answer (%) (N = 150)
Products with LC	86
Main properties of LC	
At least 1	76
At least 2	63
At least 3	59
LC as state of matter	73
LC in living organisms	61
Only 3 states of matter	76
LC molecules distribution	100

Table 2 Students' achievements on the post-test immediately after the implementation of the teaching module on hydrogels

Question about hydrogels	Percentage of students with correct answer (%) (N = 104)
Absorption of water	76
Products with hydrogels	74
Selective absorption of liquids	51
Containing water under the stress	72
Pearl as a magnifying glass	82
Some are fluorescent	76
Refractive index very similar to water	73

- *The unnecessary topic for everyone.*
- *I felt overloaded.*

Hydrogels:

- *The best thing was that we had hydrogels in our hands.*
- *Measuring with Vernier callipers was boring.*
- *Something new, finally.*
- *I told about hydrogels to my parents. We were experimenting with hydrogel from a diaper.*
- *I wanna synthesize a hair gel.*
- *I liked it.*

From data one can safely conclude that the contemporary science on liquid crystals and hydrogels might be presented in the classroom by implementing the designed teaching modules, which cover basic properties of materials, stress everyday context, and include hands-on experiments that give students personal experience with materials. It was shown that students assimilated basic knowledge and the majority of the students gave positive feedback on the modules commonly including comments to experiments. From the students' responses, it might be identified that they realized the meaning of the materials in their life. However, there are several limitations of the study referring to data collection starting from the attendance of a specific student in non-obligatory lectures, gender distribution, science interest, etc. Nevertheless, the existence of evaluated modules now enables studies on influences of lectures about contemporary science on students' attitudes towards physics.

3 Superconductivity

Superconductivity is an important context to integrate in the secondary school curricula because it constitutes a significant part of the physics of the '900 both

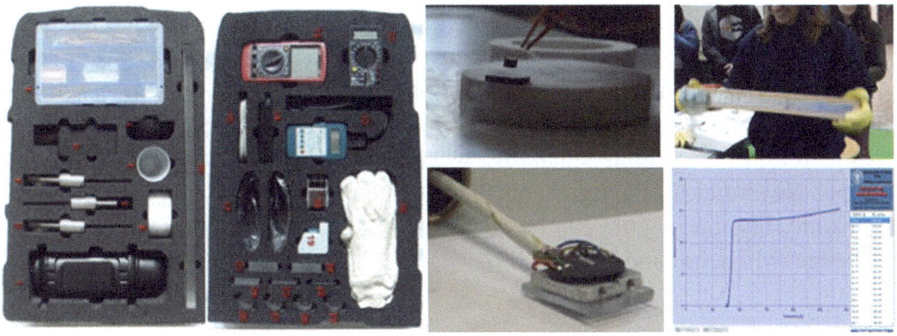

Fig. 1 High tech kit for exploring superconductivity (left) and examples of experiments (right, from the top): Meissner effect levitation of a magnet on a superconductor; model of MAGLEV train; apparatus for measurement of the breakdown of resistivity in superconductors and sample data

for theoretical implications and technological applications (Hake 2000). Different educational approaches can be adopted (Corni et al. 2009; Strehlow and Sullivan 2009; González-Jorge and Domarco 2004): a historical reconstruction of the experimental discovery and the main theoretical results; artefacts analysis of technological apparatuses based on the properties of superconductors (i.e. MAGLEV or supermagnets); explorations of the magnetic and electric properties of superconductors using simplified apparatuses. The approach developed is lacking to offer students concepts and fundamental laws of superconduction, and only a few studies consider students learning (González-Jorge and Domarco 2004; Michelini and Viola 2009). In the context of four European projects, educational materials (Fig. 1) were developed and used in Italian schools (Corni et al. 2009; Michelini and Viola 2009; Engstrom et al. 2008; Kedzierska 2010). The materials and the expertise from these projects were used in an educational project for high school on phenomena related to superconductivity (Michelini et al. 2014; Stefanel et al. 2014). Here, the outline and a few learning outcomes are presented.

3.1 *Resources and Assumptions for Educational Path on Superconductivity*

Superconductivity offers an example on how physicists explore a phenomenology without a support of a consolidated theory. In this perspective, the educational path presented in this section follows a phenomenological approach, in which students construct step by step the conceptual tools of classical electromagnetism needed to understand the superconducting state, and briefly glance at the BCS theory (Michelini et al. 2014; Stefanel et al. 2014). This approach uses a Kit, presented in Fig. 1 (Engstrom et al. 2008; Kedzierska 2010), which includes: strong neodymium magnets (0.7–0.8 T close to the poles); YBCO discs with low pinning effect, evidencing a predominant Meissner effect; high pinning effect YBCO; a model of MAGLEV train (Fig. 1a); a USB apparatus for measuring the breakdown of resistivity (Fig. 1b) (Gervasio and Michelini 2009). To overcome the well-known learning difficulties on the complex concepts of electromagnetism (Maloney 2001), the educational path follows engaging situations, where students are able to reconstruct their concepts. Nevertheless, the basic concepts of electromagnetism are prerequisite for the module, but the module can be adapted for students with weaker preliminary knowledge (Michelini and Viola 2009).

3.2 Teaching Sequence in Educational Path on Superconductivity

The learning sequence follows a logical line connecting phenomena observed and questions usually posed by students (Stefanel et al. 2014).

An YBCO disc is cooled in a liquid nitrogen with a magnet placed on it to demonstrate an evident repulsive interaction, often accompanied by a levitation of magnet over the YBCO disc. Students recognize that only the superconductor changes its properties. The repulsive force activates a spontaneous question if the YBCO disc becomes a ferromagnetic object or a magnet, when cooled. Students explore the forces approaching the magnet to the YBCO from the side, reversing the magnet or the disc, moving gently the magnet from the local equilibrium levitating position, changing the magnet, etc. The interactions remain repulsive, and the levitation always occurs even if the cooling takes place in the presence of the magnet (Meissner effect). The superconductor does not become a ferromagnetic object nor a magnet.

To characterize these magnetic properties, students can explore the interaction between a strong magnet and different systems: bars of different materials (wood, copper, aluminium, tin, brass), little vials filled with different liquids (water, alcohol) suspended on a torque balance; graphite pencil leads or pyrolytic graphite placed on a track of two additional leads; a pyrolytic graphite chip. They recognize operatively three groups of: (1) ferromagnetic materials show a strong attraction, or a rotation and then an attraction; (2) paramagnetic materials weakly attract a magnet; (3) diamagnetic materials repel it (usually very weakly). A superconductor repels a magnet and can be classified as diamagnetic. To characterize the unusual strength of interaction, students construct a sandwich composed of the YBCO disc closed between a magnet and a little iron ring. At room temperature, the magnet can lift the sandwich (the magnetic field penetrates the disc), at the liquid nitrogen temperature the magnet is unable to lift the sandwich (the magnetic field in a disc is zero or close to zero): the superconductor is an ideal diamagnetic, only in the presence of an external field and adapts to any field changes.

Diamagnetic properties suggest considering electromagnetic induction that can explain a response of a superconductor to an external magnetic field. Therefore, students experiment with eddy currents, which affect magnet falling on a metal (copper, aluminium) plate or inside metallic equal pipes (same diameter, length, thickness) from different materials (plastic; brass; aluminium; copper). They recognize that the magnet falls with a constant velocity in the metallic tubes, with different speeds for different materials, due to the time variation of the magnetic flux. Recalling Ohm's law, students recognize the role of the resistivity of different materials in these experiments. The levitation of a magnet over a superconductor can be seen as a zero speed fall down, where the electromagnetic induction occurs when the magnet approaches the superconductor and the induced current is persistent due to the absence of resistivity in a superconductor. This consideration is extended to the superconductivity in the presence of an external field (the effective Meissner effect) during the cooling. The experimental analysis of the breakdown of the resistivity of a superconductor (Gervasio and Michelini 2009) demonstrates its ideal conductive properties, explaining its diamagnetic nature. A qualitative introduction of the BCS theory gives students some basic ideas on emergence of superconductivity. The educational path ends considering the pinning phenomena and the characteristic of a MAGLEV train model.

3.3 Students' Learning Outcomes

The educational path outlined in the previous subsection was performed in 44 Italian classes. Here, we summarize a few results related to a sample of 412 students, age between 17 and 19, emerging from tutorials and pre-/post-test (Michelini et al. 2014; Stefanel et al. 2014). Facing the levitation of a magnet on a superconductor, the majority of students initially gave description like: "the cooled YBCO became a ferromagnetic" (or paramagnetic) (70%), "it became a diamagnet" (13%), "a repulsive force emerged" (20%). After exploring the properties of ordinary paramagnetic and diamagnetic systems, a large majority of students (81%) described a cooled YBCO as diamagnetic, describing the changes observed as a phase transition (68%) or simply a change in properties (15%) or in interaction behaviour (17%). Analysing the fall of a magnet inside tubes from different metals and a superconductor, students' answers are summarized in Fig. 2.

Student used magnetization vector (57%) or field lines (24%) as conceptual references. They developed models describing the condition $B = 0$ inside the superconductor (84%), assuming for their explanation an electromagnetic induction model (62%), or a magnet image model (38%).

4 Analogy Between Superconductivity and Particle Physics

In the following section, an analogy between the theory of superconductors (SC) and the weak interaction of the Standard Model of particle physics, especially the Higgs mechanism (HM), is discussed. Both theories blend in very well into a teaching

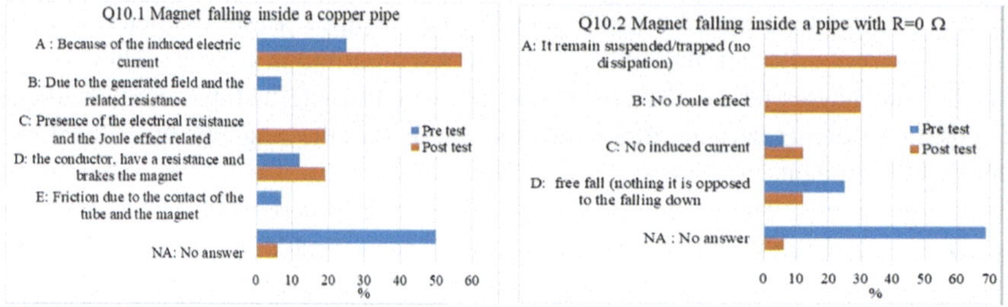

Fig. 2 Distribution of answers categories to the questions concerning a magnet falling inside a metallic tube in the pre-test and in the post-test

approach for the introduction of the basic concepts of particle physics for high school students. The teaching approach has been developed by Netzwerk Teilchenwelt. This approach focuses on the fundamental interactions of the Standard Model and the associated charges, the electric charge, the weak charge (also Isospin) and the strong charge (also colour charge). The approach is introduced to teachers all over Germany during two-day trainings that have been held in cooperation with the Dr. Hans Riegel foundation since 2016 (Lindenau and Kobel 2019). In these trainings, the HM is a highly requested topic, because the participating teachers are frequently confronted by students' questions on the matter. Although the discussion of the HM is not per se suitable for any high school course on particle physics, it could be connected to the discussion of superconductivity (Sect. 3).

4.1 *Role of Analogies*

The use of analogies has been a research field for decades and their effect on learning processes as well as possible problems and requirements for their successful use in science education can be found in literature (e.g. Duit 1991a). Analogies compare structures between two domains. By making use of the analogy, knowledge about a base domain is transferred to a target domain (Gentner and Gentner 1983). The domains can also switch positions since analogies are symmetric. Due to this symmetry, in the process of the application of the analogy, both, the target domain as well as the base domain can be developed. In addition, possible misconceptions within the base domain might be revealed during the use of analogy (Duit 1991a). Considering that, the outlined analogy might not only be used to discuss some key aspects of the weak interaction and the Higgs mechanism but also to deepen the understanding of superconductors and electromagnetism.

Analogies also play a significant role in scientific theory crafting for which the presented analogy may serve as a prime example since it has been the starting point for the development of the theory of electroweak symmetry breaking and led to the formulation of the Higgs mechanism. Considering that, the discussion of the analogy may also contribute to the learners' meta-knowledge about the nature of science and scientific research strategies and methods.

For analogies to be fruitful, Gentner (1989) identified the need for surface similarities that enable the access to the analogy and increase their acceptance. Higher-order structural similarities govern the power of the analogy and enable the students to develop sophisticated knowledge in the target domain (Gentner and Landers 1985; Holyoak and Koh 1987). Furthermore, a sophisticated understanding of the base domain is required for the analogy to be useful. This is the main reason why the discussed analogy is not suitable for any high school course, as stated above. In this context, it's worth mentioning that an introduction to superconductivity can also be a useful addition to a course on particle physics, since it is a key technology for high energy particle physics research as carried out at CERN.

4.2 *Outline of Analogy and Connection to the General Teaching Approach*

The major difference between the weak and the electromagnetic interaction of the Standard Model of particle physics is, that the weak interaction has a very short effective reach. This means that the associated force falls off quasi-exponentially at distances in the order of 0.001 fm. Trying to visualize such a force law with the field line model leads to a conflict that can be used to introduce the messenger particle model (Lindenau and Kobel 2019). Figure 3 shows a possibility of the visualization of field lines of the weak force.

Since the force diminishes very quickly, the field line density has to drop rapidly and field lines need to come to nothing. In Fig. 3 the lines vanish stepwise to indicate a quasi-exponential fall off of the force. This behaviour is not compatible with the rules

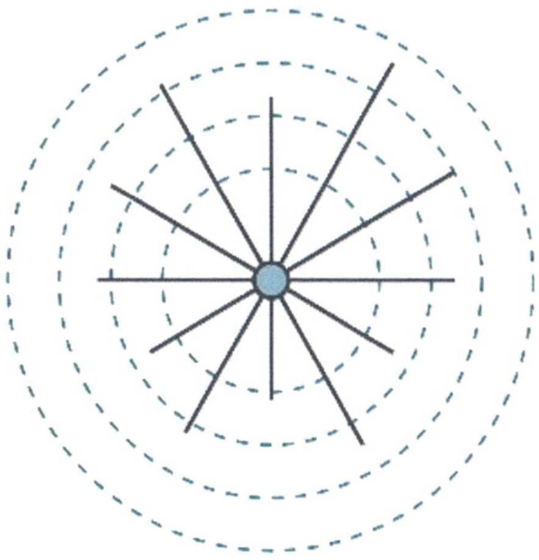

Fig. 3 Disappearing field lines for the weak force (Lindenau and Kobel 2019)

of the field line model in vacuum. But there are materials that students are familiar with and that effect the electric field strength, namely dielectrics. With the knowledge of superconductors (SC) students even know materials in which any external electric and magnetic field completely vanishes. More precisely, the fields can penetrate the SC for a small distance, the so-called London penetration depth while it falls off exponentially exactly as the weak force does. This behaviour of a rapid decreasing of a field and therefore the measurable force is the surface similarity between SC and the HM, which in the messenger particle model leads to the large masses of the W and Z particles. The Higgs field as an omnipresent field possess weak charge itself and therefore is responsible for the screening of any weak charge of elementary particles.

Table 3 shows a (not complete) list of objects and processes that map from the Ginsburg-Landau (GL) model as well as from the Bardeen-Cooper-Schrieffer (BCS) model of superconductivity to the HM (Dixon 1996; Fraser and Koberinski 2016). The mappings can be considered to be structural similarities. Depending on the scope of the discussion, fewer or even more similarities might be addressed.

As with any analogy, there are limitations that should be discussed in the discourse. It is important to point out that the HM has more structural similarities with the GL model than with the BCS model. Similar to the GL model, the HM is a phenomenological description of the electroweak symmetry breaking. That means, there is no causal explanation of the process of symmetry breaking and how the phase transition (the condensation of the Higgs field) can be explained. Therefore, within the HM no object can be identified which takes the role of the Cooper pairs and there is no validated explanation similar to the description of the development of the Cooper pair condensate in the BCS theory. Discussing those limitations of the analogy can be a chance to draw the learners' attention to open research questions in the field of particle physics and to point out that with the discovery of the Higgs particle at

Table 3 Some structural similarities between SC theory and the HM

SC theory	Higgs mechanism
Superconducting material	Universe
Combined wave function of superconducting electrons (GL)	Higgs field
Cooper pair condensate (BCS)	Higgs field
Electric charge number −2	Weak charge number −1/2
London penetration depth	Compton wavelength of the W particle
Density fluctuations of Cooper pair condensate (observed 2013)	Higgs particle
Coherence length of the SC	Compton wavelength of Higgs particle

CERN in 2012 the experimental testing of the Higgs mechanism has not ended but actually just begun.

4.3 Possible Extensions to the Analogy

There are several extensions that can be made to the analogy that might support learning. For example, further bridging analogies (Brown and Clement 1989) can be included. An example is a discussion of diamagnetics in addition to dielectrics before comparing the weak interaction with the theory of superconductivity. Inclusion of metaphors can help students to develop the analogy themselves while connecting affective and cognitive domains due to their abnormal aspects (Duit 1991b). An example might be the following sentence: "For photons, being inside of a superconductor is like a visit at Burger King". In this metaphor, it is addressed that photons within a SC can be described as massive particles in the same way that W particles are always massive due to their interaction with the omnipresent Higgs field, which causes them to have a finite reach. That is obviously also true for photons within a superconductor.

Not mentioned here are similarities concerning the symmetry breaking (like order parameters). In this context, the symmetry breaking happening in ferromagnetic materials below the Curie Temperature can serve as a bridging analogy. Furthermore, additional similarities in the mathematical structure (e.g. the form of the potentials) have been left out. The role of the Higgs particle as an excitation of the Higgs field has not been discussed either, but it has at least to be pointed out that the Higgs field is the relevant subject of the analogy and the Higgs particle experimentally only serves as an object of investigation to learn more about the properties of the associated field itself.

5 Professional Development Programmes in Particle Physics: A Delphi Study

As described in the previous sections, scientific knowledge is rapidly expanding. Discoveries and new technologies are making their way into the classroom. However, these fast developments can sometimes go beyond the teachers' initial training, and teachers might find themselves feeling lost. Therefore, teachers around the world are looking for professional developments that will help them advance their skills and knowledge (Greene et al. 2013; Hewson 2007; OECD 2019). More than 95% of the teachers in the OECD countries reported participation in professional development in 2018 (OECD 2019). Consequently, many large research institutions started developing professional development programmes. Among them is CERN, the European

Organisation for Nuclear Research and the world's largest particle physics laboratory. As the source of many recent developments in science and technology, CERN is the perfect place for educating teachers on contemporary topics, especially in particle physics.

CERN's Teacher Programmes are open to in-service high school science teachers from around the world and welcome around one thousand teachers every year. Specifically, CERN offers two types of programmes, national programmes in the teachers' national language, lasting between three to five days, and two-week international programmes, which are delivered in English. Both types of programmes contain a series of lectures by experts in modern science and technology related to CERN, guided tours to on-site research facilities, hands-on workshops, and social events. As the programmes can differ in content and length, we wanted to determine which topics in modern science and technology connected to CERN should be included in the different CERN's Teacher Programmes. Therefore, we conducted a study to investigate which topics relevant stakeholders believe are most important for the teachers.

5.1 Delphi Study Design

To elicit expert opinions without being much affected by normative social influence, we needed to find a structure that allows anonymous discussions without face-to-face interactions (Osborne et al. 2003b; Rowe and Wright 1999). Therefore, we used the design of a conventional Delphi study, which provides a robust iterative structure for a detailed critical anonymous discussion between experts. In our study design, the experts answered three consecutive questionnaires with interspersed feedback, which allowed deliberations on the results of the previous questionnaire before filling out the subsequent one. The experts were selected through nominations. They formed five panels: (1) Physics education researchers with experience in organizing and/or investigating professional development programmes, (2) CERN national coordinators, who help organizing CERN's national teacher programmes, (3) Members of the CERN council and advisory boards (short: CERN management) with high knowledge of CERN, (4) Teachers, who participated in CERN's Teacher Programs in the past, and (5) Teachers, who have applied to participate in the future. Here, the last panel was only included in the second round of the study. The overall number of participants by the panel and by round are presented in Table 4.

The first round open-ended questionnaire was designed to elicit experts' opinions and ideas on the goals, objectives, and design features of professional development programmes in general as well as at CERN and similar large research institutions. The results were analysed with inductive thematic analysis. Here, we built themes and overarching categories based on the experts' responses. The results were compiled into a feedback summary and sent to the participants for deliberation. Additionally, the emergent themes were used as the basis for the subsequent questionnaires.

Table 4 Expert participants in the three rounds of the Delphi study

Panel	1st round	2nd round	3rd round
Physics education researchers	28	26	31
CERN national contacts	24	20	14
CERN management	16	10	10
Teachers: past participants	13	28	17
Teachers: future participants	–	16	18

In the second round questionnaire, the experts rated the themes on a 6-scale Likert-like importance scale, ranging from "Very unimportant" to "Very important". The results from this round were quantitatively analysed using the Likert package and the Kruskal-Wallis test. Additionally, we also qualitatively analysed the comments of the experts. Again, the results of the analysis were summarized and communicated to the experts, together with the third round questionnaire. As the overall outcomes of the second round showed a strong ceiling effect, we included ranking questions in the third round questionnaire. Here, we merged some themes with the help of three additional experts in particle physics and physics education to reduce the burden on the experts (Kranjc Horvat et al. 2019).

The results of the third round were analysed using the Kruskal-Wallis test with Bonferroni adjustment to compare the rankings of the respective panels of experts. Additionally, we analysed the distances between the rankings. Due to overlapping, we grouped the topics based on their medians. Here, all topics with median rank higher than the interquartile range of the entire ranking were grouped in "high", topics inside the interquartile range "medium", and topics ranked lower, "low". The groups were then re-tested to ensure that the group ratings are significantly different.

Additionally, the experts were given forty hours, which they had to distribute among the various themes, judging by how many hours they think should be devoted to each of the themes in a teacher training programme. Using the correlation test, we determined the level of correlation between the ranking and the hour assignment. Furthermore, the experts' comments were qualitatively analysed using inductive thematic analysis to extract any possible missing themes and to enhance the interpretation of the quantitative results.

5.2 Results

The inductive thematic analysis of the results of the first round questionnaire provided us with seven overarching categories, including over one hundred themes. However, to stay within the scope of this paper, we will only focus on the category "Topics at

CERN and similar large research institutions". Here, we were able to identify seventeen major themes, which were then included in the subsequent questionnaire. In the second round, the experts were asked to rate each of them based on their perceived importance. The results of the second round analysis showed a prominent ceiling effect, with all topics being recognized as important by more than 75% of experts. Additionally, no significant differences could be identified between the panels.

The third round questionnaire was designed to allow experts to rank the themes based on their perceived importance. Some themes were merged to increase the reliability of the study, bringing the total number of themes down to thirteen. Additionally, we added "Other" as the last theme to allow for new ideas from the experts. The analysis using the Kruskal-Wallis test showed no significant differences between the rankings of different panels in general ($p > 0.05$). Therefore, it is safe to say that the different panels generally agree on the ranking. However, the same test also showed that there are minimal differences between the ranks of different items. Therefore, we grouped the items into three groups, based on their perceived importance (high, medium, and low), as shown in Table 5. The difference between such groups is significant ($p < 0.005$).

Additionally, the experts were asked to assign hours to each of the topics. Here, they were given a total of forty hours, which they had to redistribute among the topics. The distribution is based on how many hours they believe should be given to each of the topics in a regular programme. The analysis with Kendall's correlation test showed weak correlation ($R = 0.34$) between the ranking and the allocated hours. However, the correlation coefficient is higher ($R = 0.44$) when comparing the grouped rankings with the allocated hours. Therefore, the more important topics

Table 5 Results of the ranking and hour allocation in the third round questionnaire in the category "Topics". Both ranks and the allocated hours represent the expert group averages

Rank	Topic	Allocated hours
High	The development and the mysteries of the Universe	4.1
High	The standard model of particle physics	3.3
High	Real-life applications of research	4.5
Medium	The latest findings and newest results	2.5
Medium	Particle accelerators and how they work	2.7
Medium	Antimatter research	4.4
Medium	Particle detectors and how they work	2.5
Medium	The technology used in modern scientific research	3.3
Medium	Einstein's theory of relativity	3.5
Medium	Data analysis in modern research (e.g. search for Higgs)	2.3
Medium	Philosophy of science and societal impacts of science	1.8
Low	History of science	2.1
Low	Advances in theoretical physics (e.g. supersymmetry)	1.9
Low	Other	1.1

have been assigned more hours than the topics that are perceived as less important, as shown in Table 5.

5.3 Discussion and Conclusion

The study on the important topics within CERN's Teacher Programmes provided a very promising overview of how important are different topics that are to be included in the programmes. Here, the experts recognized as the most important the following topics: the development and the mysteries of the Universe, real-life applications of research, and the Standard Model of particle physics. From the experts' comments, we can see that all topics that were rated high were perceived as more important either due to their connection to curriculum (e.g. the development of the Universe and the Standard Model) or their perceived higher appeal to students (e.g. the development of the Universe and real-life applications). Therefore, the programmes should focus mostly on the topics that can be easily connected to the existing curriculum.

Another important outcome of the study comes from the analysis of the correlation between the ranking of the topics by importance and the number of assigned hours. Here, the study shows that, in general, topics that are perceived as more important should get more dedicated hours. Although we cannot claim that importance is the only factor in the number of hours assigned, its influence should not be ignored. Still, further studies are needed to determine all the influencing factors and their respective impact on the number of hours assigned to a topic.

On the other part of the scale, we find topics that were perceived as harder to connect to the existing curricula and, as such, less important. Consequently, the number of hours for these topics is also lower. However, all of the topics in the list were rated important by a majority of the experts in the second round of the study. Therefore, lower-rated topics should not necessarily be excluded out of the programmes. Furthermore, with close to only one hour assigned to the category "Other", the list of the topics to be included in CERN's Teacher Programmes is most likely complete and provides a sound basis for organizing future programmes.

6 Conclusions

At a time, when the positive attitude towards mathematics, science and technology at school seems to start to drop already at early ages (between ages 8 and 11) (Sokołowska et al. 2015) we search for diverse solutions to increase pupils' interest, motivation and engagement in these subjects. Some EU countries [e.g. Belgium, see (https://onderwijs.vlaanderen.be/nl/wat-is-het-stem-actieplan)] have already started campaigns for reinforcement of STEM education, but still in most of the curricula across Europe, especially for physics, the set of school topics is traditional and no space is secured for students' interest on contemporary discoveries or novel applications. If at all, such topics are usually introduced only in an informative way. However, it was proved that introduction of a contemporary topic to the high school programme through hands-on experiments increases students' positive attitude and strengthen motivation to continue learning about this topic (Blonder and Dinur 2011). Since only few examples of such learning units incorporating current scientific hot issues can be found in the literature (Čepič 2017), and the research on their implementation is even more limited, there emerges a considerable need in this area to be addressed.

The symposium presented in this paper provided a few good examples of the introduction of contemporary scientific issues to physics classes at the pre-university level. Three ready-to-use learning units on liquid crystals, hydrogels and superconductivity were presented together with the study of their implementation researching the impact on students' acquirement of the knowledge and their perception of the implementation. The results showed a significant gain of the content knowledge and a considerable development of basic concepts related to these three topics. Most of the students taking part in soft matter laboratories (liquid crystals, hydrogels) enthusiastically summed up their participation in the activities, indicating the motivational value of the novelty of the topics and an engaging format of the experimenting by themselves.

For more advanced students a proposal of a unit on the analogy between superconductivity and particle physics was presented, showing the strength of analogies in physics education and in professional physics research.

A design of the learning units based on contemporary topics is only the first necessary step that needs to be undertaken together by researchers and educators, experts in two complementary fields—advanced physics and physics didactics. However, a successful implementation requires the most important actor, a teacher who will bring the idea to the classroom. Most of the in-service teachers feel like aliens when we talk about contemporary research issues that in many cases were not even known at the moment when the teachers studied. Thus appropriately tailored teacher training is an indispensable part of the entire module design process. One of very good examples of such an approach to physics education is work done by CERN through professional development programmes and described in the last contribution. Since at CERN teachers are trained by experts in different fields it occurred necessary to gather their opinions about the themes that need to be addressed during such trainings. Thus, a three-rounds Delphi study was implemented that helped to rank the topics to be covered in shorter and longer courses for teachers at CERN.

It seems that incorporating contemporary topics into the physics curriculum can be interesting and motivating for students, but at the same time challenging for their teachers. Thus successful implementation should combine four steps: (1) preparation of the hands-on module in collaboration between researchers in the field and educators, (2) practical teacher training based on the module together with research on teachers' opinions, (3) research on implementation in the classroom, including a study on development of students' concepts, gain of the content knowledge, research skills development and students' perception of the implementation, (4) redesign of the unit on the basis of the results from (2) and (3). But it is worth the effort, since it brings a new quality to the curriculum that so urgently needs to be transformed in a time when the physics subject at school lags behind the reality experienced anywhere else.

Acknowledgements The financial support of Slovenian Agency for research and development, the project J7-8278, is acknowledged.

References

Blonder R, Dinur M (2011) Teaching nanotechnology using student-centered pedagogy for increasing students' continuing motivation. J Nano Ed 3:51–61. https://doi.org/10.1166/jne.2011.1016

Brown DE, Clement J (1989) Overcoming misconceptions via analogical reasoning: Abstract transfer versus explanatory model construction. Instr Sci 18:237–261. https://doi.org/10.1007/BF00118013

Calo E, Khutoryanskiy VV (2015) Biomedical applications of hydrogels: a review of patents and commercial products. Eur Polym J 65:252–267. https://doi.org/10.1016/j.eurpolymj.2014.11.024

Čepič M (2017) Introduction of current scientific results to education: experiences from the case of liquid crystals Key competences in physics teaching and learning: selected contributions. In: Greczylo T, Debowska E (eds) The international conference GIREP EPEC 2015, Wrocław Poland, 6–10 July 2015. Springer Proceedings in Physics, 190, pp 41–54

Corni F, Michelini M et al (2009) http://lsg.ucy.ac.cy/girep2008/intro.htm

Dixon L (1996) From superconductors to supercolliders. Beam Line 26:23–30

Duit R (1991a) On the role of analogies and metaphors in learning science. Sci Educ 75:649–672. https://doi.org/10.1002/sce.3730750606

Duit R (1991b) Students' conceptual frameworks: consequences for learning science. Psychol Learn Sci 75:649–672

Dunmur D, Sluckin T (2011) Soap, science and flat-screen TVs, a history of liquid crystals. Oxford University Press Inc., New York

Engstrom V, Ireson G, Latal H, Mathelisch L, Michelini M, Peeters W, Rath G (2008) In: Sidharth BG et al (eds) The Supercomet2 Project FFP9. AIP, New York, vol 1018, pp 221–224

Fraser D, Koberinski A (2016) The Higgs mechanism and superconductivity: a case study of formal analogies. Stud Hist Philos Sci Part b: Stud Hist Philos Mod Phys 55:72–91. https://doi.org/10.1016/j.shpsb.2016.08.003

Gentner D (1989) Analogical learning similarity and analogical reasoning. Cambridge University Press, Cambridge, pp 199–244

Gentner D, Gentner DR (1983) Flowing waters or teeming crowds Mental Models. Erlbaum, Hillsdale, pp 99–129

Gentner D, Landers R (1985) Analogical reminding: a good match is hard to find. In: Proceedings of the international conference on systems, man and cybernetics

Gervasio M, Michelini M (2009) http://www.fisica.uniud.it/URDF/mptl14/contents.htm

González-Jorge H, Domarco G (2004) Superconducting cylinders aid in an understanding of current induction. Phys Educ 39:234. https://doi.org/10.1088/0031-9120/39/3/F04

Greene BA, Lubin IA, Slater JL, Walden SE (2013) Mapping changes in science teachers' content knowledge: concept maps and authentic professional development. J. Sci Ed Technol 22:287–299. https://doi.org/10.1007/s10956-012-9393-9

Hake RR (2000) Is it finally time to implement curriculums? AAPT Announcer 30(4):103

Hewson PW (2007) Teacher professional development in science. Handb Res Sci Educ 1:1179–1203

Holyoak KJ, Koh K (1987) Surface and structural similarity in analogical transfer. Memory Cogn 15:332–340

https://onderwijs.vlaanderen.be/nl/wat-is-het-stem-actieplan (on-line translation available)

Kedzierska E et al (2010) MOSEM 2 project. Il Nuovo Cimento 33C(3):64–74

Kranjc Horvat A, Wiener J, Schmeling S, Borowski A (2019) Goals and objectives of teacher training programmes: lessons learned from a multi-stakeholder analysis. Nastava Fizike 9:1–10

Lindenau P, Kobel M (2019) Introducing an innovative approach of teaching the Standard Model of particle physics at high school. J Phys: Conf Ser 1287:012045. https://doi.org/10.1088/1742-6596/1287/1/012045

Maloney DP et al (2001) Am J Phys 69(7):S12–S23

Michelini M, Viola R (2009) In: Constantinou CP, Papadouris N (eds) Physics curriculum design, development and validation. GIREP, Cyprus. http://lsg.ucy.ac.cy/girep2008/intro.htm

Michelini M, Santi L, Stefanel A (2014) In: Burra G, Michelini M, Santi L (eds) Basic concept of superconductivity FFFP12. Springer, Cham, pp 453–460

OECD (2019) TALIS 2018 results (Volume 1): Teachers and school leaders as lifelong learners, TALIS

Okay O (2009) Gerlach G, Arndt KF (eds) General properties of hydrogels: hydrogel sensors and actuators. Springer, Berlin, pp 1–14

Osborne J, Simon S, Collins S (2003a) Attitudes towards science: a review of the literature and its implications. Int J Sci Ed 25:1049–1079. https://doi.org/10.1080/0950069032000032199

Osborne J, Collins S, Ratcliffe M, Millar R, Duschl R (2003b) What "ideas-about-science" should be taught in school science? A Delphi technique of the expert community. J Res Sci Teach 40:692–720. https://doi.org/10.1002/tea.10105

Pavlin J, Čepič M (2017) Hydrogels in the classroom key competences in physics teaching and learning: selected contributions. In: Greczylo T, Debowska E (eds) The international conference GIREP EPEC 2015, vol 190, Wrocław Poland, 6–10 July 2015. Springer Proceedings in Physics, pp 191–201

Pavlin J, Vaupotič N, Čepič M (2013) Liquid crystals: a new topic in physics for undergraduates. Eur J Phys 34:745–761. https://doi.org/10.1088/0143-0807/34/3/745

Rowe G, Wright G (1999) The Delphi technique as a forecasting tool: issues and analysis. Int. J. Forecasting 15:353–375

Sokołowska D, de Meyere J, Wojtaszek M, Zawadzki W, Brzezinka G (2015) Attitude, motivation and self-esteem in mathematics, science and technology researched by the SECURE project Polish results against the background in Europe. Edukacja Biologiczna i Środowiskowa 1(2015):27–38

Stefanel A, Michelini M, Santi L (2014) In: Tasar F (ed) High school students analyzing the phenomenology of superconductivity WCPE 2012. Pegem Akademiel, Ankara, pp 1253–1266

Strehlow CP, Sullivan MC (2009) A classroom demonstration of levitation. Am J Phys 77:847–851. https://doi.org/10.1119/1.3095809

Roland Eötvös and the Equivalence Principle

Roland Eötvös: Scientist, Statesman and Educator

András Patkós

Abstract This lecture recalls the memory of Baron Roland Eötvös, an outstanding figure of the experimental exploration of the gravitational interaction and "funding father" of applied geophysics. Beyond the scientific achievements, his contribution to the development of the modern Hungarian schooling and higher educational system, most importantly, the foundation of an innovative institution of teacher's training did not lose its contemporary significance. This lecture has been invited by the organizers of this conference in response to the decision of UNESCO to commemorate world-wide the death centenary of the most outstanding Hungarian experimental physicist of modern times.

Keywords Eötvös rule · Eötvös balance · Eötvös effect · Joseph Eötvös College

1 Curriculum Vitae (Dobszay et al. 2019)

The noble family Eötvös has played an important role in the nineteenth-century history of Hungary. József (Joseph) Eötvös was a prominent political leader of the social reform period before the 1848 revolution and became minister of education and religious cults in the first independent Hungarian Government. After the radicalization of the revolution, he has spent a short period in exile. Later, he has participated actively in the political negotiations leading to the formation of the Austro-Hungarian Monarchy in 1867, after which once again he entered the Hungarian Government.

József Eötvös, though he has recognized early the affection of his son, Loránd (Roland) to investigating natural phenomena, still has convinced him to study for two years law at University of Pesth before joining University of Heidelberg. He wrote to his son in 1866: "If your soul, tired in sciences after a few years, would aim at activities on a wider horizon, I could not justify myself depriving you of the necessary basic legal education."

A. Patkós (✉)
Eötvös University, Budapest, Hungary

B. Jarosievitz and C. Sükösd (eds.), *Teaching-Learning Contemporary Physics*, Challenges in Physics Education,
https://doi.org/10.1007/978-3-030-78720-2_6

The academic and public carrier of Roland Eötvös took a very steep start after he has received the doctoral degree in Heidelberg as one can see from Table 1 next page. There is a short list of the important events of his life for the curious reader which is encompassed in the second column by his age. As a highly respected scientist, he served in several positions at University of Budapest and has been elected as the ever-youngest president of the Hungarian Academy of Sciences. He has initiated the foundation of the Mathematical and Physical Society and of the Hungarian Tourist Association. He acted for a short time as Minister of Cults and Education. During his ministerial commitment, he has finalized the laws ensuring religious freedom and has enlarged the network of the "elementary" schools. With the foundation of the József Eötvös College and earlier of the Mathematical and Physical Society, he has contributed innovatively at improving the formation and the working conditions of high-school teachers.

Starting with the Paris meeting of the International Conference of Surveyors in 1900, he regularly has presented results obtained with the famous Eötvös balance and has received increasing international attention at the subsequent Budapest (1906) and Hamburg (1912) meetings. He has been nominated to the 1911 Nobel prize (received eventually by Kammerlingh-Onnes).

On the occasion of his 70th birthday (1918), the Mathematical and Physical Society has published a volume reviewing his scientific and educational achievements, which has been reedited in 1930 by the Hungarian Academy of Sciences Fröhlich (1930). The Geophysical Institute founded by himself has taken his name directly after his death in 1921. Also, the Mathematical and Physical Society has been named after him (1923) as well as a college supporting outstanding students of poor social origin, originally initiated by the famous physicist Zoltán Bay, a former Eötvös-collegiate (1929). Since then, many streets, schools and institutions bear his name in Hungary, including the University of Budapest (1950).

Below, in Sect. 2, I shall discuss his scientific discoveries which preserve a place for his name in many monographies of chemistry, physics and geophysics. In Sect. 3, a brief summary is presented of his views and actions on improving the quality of high-school teaching.

2 Three Important Discoveries Bearing the Name of Roland Eötvös

2.1 *Accurate Experimental Method for the Determination of the Capillary Constant and the Eötvös-Rule of Its Temperature Dependence (Eötvös 1886)*

In 1869, Eötvös has attended the lectures of Franz Neumann in Königsberg who also gave lectures on his theory of capillarity. The fluid wets and "climbs" up along the walls of a container as is shown in Fig. 1 for the case of a liquid between two

Table 1 Most important events of the life of Roland Eötvös

Year	Age	Family/Private	Scientific&Academic	Public&Social
1848		Born in *Buda*		
1868-70	20-22	PhD Studies in *Heidelberg*		
1871	23		Habilitation at University of *Pesth*	
1873	25		Corresponding member of Hungarian Academy of Sciences	Editor of the science popularization monthly Communications in Natural Sciences
1876	28	Married on Gizella Horvát (2 daughters)	**New method for the experimental determination of the *capillary constant***	
1878	30		Head of the Department for Experimental Nature Studies at University of *Budapest*	
1880	32	Yearly summer mountaineering till 1914		Vice-President of the Society for Advancement of Natural Sciences (till 1919)
1886	38		**Eötvös-rule of the *temperature dependence* of the capillary constant**	
1888-90	40-42		**New method for *testing the law of Universal Free Fall***	1889: President of the Hungarian Academy of Sciences (till 1905)
1891	43		**Construction of the *Eötvös-balance***	**Foundation of the Mathematical and Physical Society (president till 1919)**
1892-95	44-47		1892-93: Rector of the University of Budapest	Minister of Education and Cults (1894-95)
1895	47			**Foundation of the József Eötvös Collegium (curator till 1919)**
1906-1909	58-61	1902: Eötvös peak in Dolomites	***Improved bound* on the validity of Universal Free Fall**	
1913	65		**Laboratory demonstration of the *Eötvös-effect***	
1919	71	Died in Budapest		

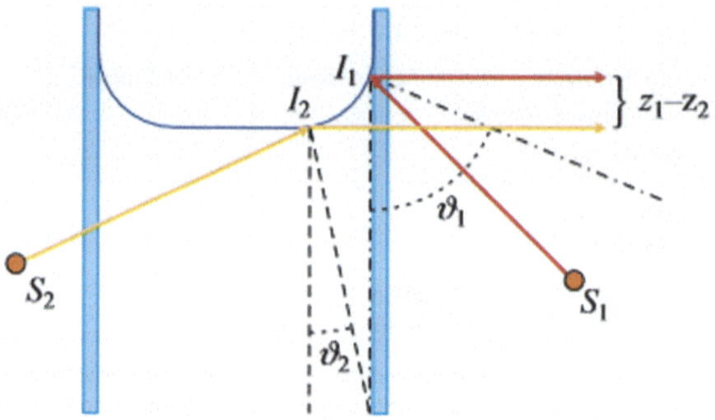

Fig. 1 Direction of the light incidence is adjusted in such way that from the source located at S_1 and S_2; it is reflected at points I_1 and I_2 exactly horizontally into the telescopes located at heights z_1 and z_2, respectively. Measuring the corresponding angles ϑ_1 and ϑ_2 one can compute the surface tension with the relation given in the text

parallel glass plates. The height (z_l) of a surface point (I_l) measured relative to the horizontal surface of the fluid far from the wall is related to the angle between the external normal and the vertical directions (ϑ_l) of a ray of light reflected exactly horizontally at the point through a simple functional form containing the capillary constant α hidden in a constant a containing also the gravitational acceleration g and the density ρ of the liquid. The relation is determined by the equation proposed independently by Gauss, Laplace and others for the determination of the curvature radius R of the liquid surface:

$$\frac{1}{R} = \frac{2z}{a^2}, \quad a = \frac{2\alpha}{g\rho}. \tag{1}$$

Student Eötvös was praised by his professor for having proposed a method for measuring the capillary constant by determining the height difference of two surface points and the corresponding angles.

Eötvös has solved the differential equation arising when one expresses the radius of curvature through the derivative of the $z(x)$ function where x is the horizontal distance of the point I measured from the wall:

$$R = \frac{\left(1 + z'^2\right)^{\frac{3}{2}}}{z''}, \quad z' = \frac{\mathrm{d}z}{\mathrm{d}x}. \tag{2}$$

From the solution $\frac{z^2}{a^2} = 2(\sin\frac{\vartheta}{2})^2$, one easily forms the convenient difference which allows the accurate determination of the combination a:

$$z_1 - z_2 = \sqrt{2}a\left(\sin\frac{\vartheta_1}{2} - \sin\frac{\vartheta_2}{2}\right). \tag{3}$$

Eötvös has published his method only in 1876. His superior method allowed him to pursue a rather extended investigation of the temperature dependence of the surface tension.

Following the derivation of the general gas law based on the hypothesis of molecular substructure by van der Waals, he has derived an analogous law for the surface tension (Eö is the Eötvös-constant):

$$\alpha V^{\frac{2}{3}} = E\ddot{o}\,(T_* - T). \tag{4}$$

The structural analogy with the formula pV = RT is obvious. The Eötvös-constant plays the same role as the Regnault-constant R.

His student of the epoch, Professor Károly Tangl, wrote in 1930: "Whenever the theoretical result has been formulated, feverish laboratory work has been started for the verification of the rule. The windows of the old physics building of the university have remained light frequently in the late nights." Eötvös and his collaborators have checked the universal nature of his law on a large number of materials before he has presented his results in 1885 at a session of the Academy of Sciences. He has identified $\boldsymbol{T}_*$ with the critical temperature, where the surface tension vanishes. This has been corrected later by Ramsay to lie somewhat lower.

An important trait of the scientist's character can be observed already in this early story. His deductions were always based on experimental facts, obtained with a procedure as perfect as the measurement methods of the epoch have allowed to reach. He did not publish intermediate "progress reports," and only the final and best results were presented to the scientific community.

2.2 Measurement of the Local Variation of the Gravitational Potential with the Eötvös Balance and Checking Universal Proportionality of the Inertial and Gravitational Masses with Unprecedented Accuracy (Eötvös et al. 1922)

The scientific challenge of his life has met Eötvös at age 40. This was the Definition I of the Principia of Newton which states the universal proportionality of the inertial mass ("resisting" to the accelerating force) and the gravitational mass (deduced from the weight of a body), which Newton claims to have checked by the most accurate pendulum measurements (see Fig. 2).

In 1890, when Eötvös has presented results from the first series of experiments checking this proportionality with 1:20.000.000 accuracy, he gave the following natural motivation for performing the experiments: "Logics require to verify the fundamental statement with the accuracy one can measure the weight of a body." By the middle of the XIXth century, weights of bodies could have been measured with $1{:}10^6$ maximal error, though the universal nature of the proportionality has been

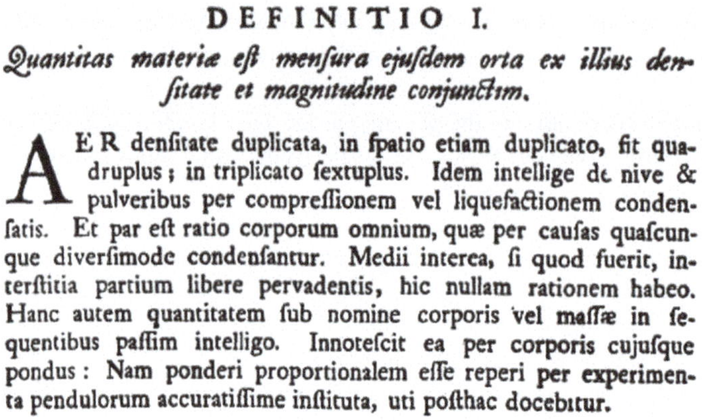

DEFINITIO I.

Quantitas materiæ eſt menſura ejuſdem orta ex illius denſitate et magnitudine conjunctim.

AER denſitate duplicata, in ſpatio etiam duplicato, fit quadruplus; in triplicato ſextuplus. Idem intellige de nive & pulveribus per compreſſionem vel liquefactionem condenſatis. Et par eſt ratio corporum omnium, quæ per cauſas quaſcunque diverſimode condenſantur. Medii interea, ſi quod fuerit, interſtitia partium libere pervadentis, hic nullam rationem habeo. Hanc autem quantitatem ſub nomine corporis vel maſſæ in ſequentibus paſſim intelligo. Innoteſcit ea per corporis cujuſque pondus: Nam ponderi proportionalem eſſe reperi per experimenta pendulorum accuratiſſime inſtituta, uti poſthac docebitur.

Fig. 2 Fundamental statement of *Philosophiae naturalis principia mathematica of Newton*

checked with help of the so-called reversible pendulum developed by F. Bessel in 1830 only to $2{:}10^5$.

Eötvös has replaced the pendulum by the Cavendish-Coulomb torsion balance and reached 1:20.000.000 accuracy. The working principle of the equipment can be simply derived with help of Fig. 3, left.

One orients ideally the torsion fiber vertically (parallel to the local direction of the full gravitational acceleration in the point of hanging a beam of length $2a$, joining two point-like bodies of mass m placed to the two ends). By the inhomogeneity of the gravitational potential, the following force components act in the horizontal (x, y)-plane in the point $(\Delta x, \Delta y)$ when the gradient of the gravitational potential energy is expanded to first order in the small coordinates:

$$F_x = \rho(0)(U_{xx}(0)\Delta x + U_{xy}(0)\Delta y),$$
$$F_y = \rho(0)(U_{yx}(0)\Delta x + U_{yy}(0)\Delta y). \quad (5)$$

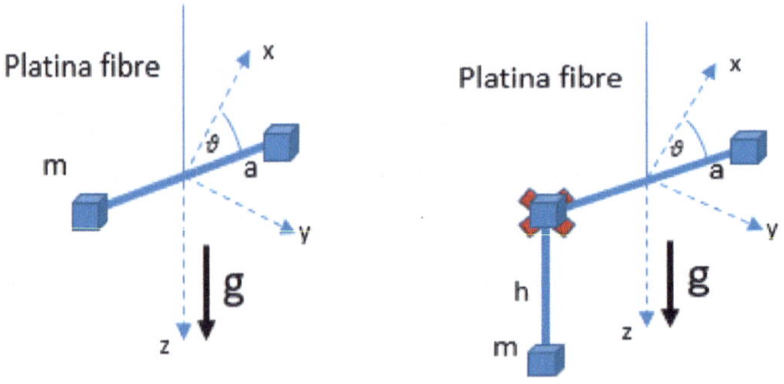

Fig. 3 Working principle of the Cavndish-Coulomb torsion balance (left) and of the Eötvös balance (right)

Here, $V(\boldsymbol{x}) = \rho(\boldsymbol{x})U(\boldsymbol{x})$ is the gravitational potential energy density for a local mass density $\rho(\boldsymbol{x})$, and U_{ij} is the second derivative of the gravitational potential with respect to the coordinates x_i and x_j. Substituting these expressions into the formula giving the torque along the z-direction one finds:

$$M_z = \int\left(\Delta x F_y - \Delta y F_x\right) = \left(\theta_{xx} - \theta_{yy}\right)U_{xy}(0) + \theta_{xy}\left(U_{yy}(0) - U_{xx}(0)\right), \tag{6}$$

where θ_{ij} is the components of the inertial momentum tensor. In the above geometry: $\theta_{xy} = \mathrm{ma}^2 \sin\vartheta\cos\vartheta$, $\theta_{xx} - \theta_{yy} = \mathrm{ma}^2[(\cos\vartheta)^2 - (\sin\vartheta)^2]$.

In 1891, Eötvös has designed a modified balance where the weight at one end is lowered by an amount h (Fig. 3, right). The advantage of this geometry is that it is sensitive also to two further components of the second derivative tensor of the gravitational potential. The additional torque is given as

$$\begin{aligned}\Delta M_z &= a\cos\vartheta\,\Delta F_y - a\sin\vartheta\,\Delta F_x \\ &= \mathrm{ma}\left[h\left(\cos\vartheta\frac{\partial g}{\partial y} - \sin\vartheta\frac{\partial g}{\partial x}\right)\right] \\ &= \mathrm{mah}(U_{zy}\cos\vartheta - U_{zx}\sin\vartheta).\end{aligned} \tag{7}$$

Now, there is a third contribution if the two masses of different material constitution experience different gravitational force, e.g., different values of the Newton's constant: $G' = G_0(1+\kappa_1)\quad G = G_0(1+\kappa_2)$, while the centrifugal force proportional to the inertial masses is identical for the two: $-\mathrm{ma}G_0\sin\vartheta(\kappa_1 - \kappa_2)\sin\varepsilon$. Here, ε is the small angle between the direction of the weight vector and the (average) gravitational force. Its value is nearly 6° at the latitude of Budapest. The equation defining the change in the angular equilibrium direction of the beam relative to the torque-less case, $\varphi - \varphi_0$ has the following form (τ is the torsion constant of the fiber):

$$\tau(\varphi - \varphi_0) = \mathrm{K}\left[U_{xy}(0)\cos 2\vartheta + \left(U_{yy}(0) - U_{xx}(0)\right)\frac{\sin 2\vartheta}{2}\right] + \mathrm{mah}\left(U_{zy}\cos\vartheta - U_{zx}\sin\vartheta\right) - \mathrm{ma}G_0\sin\vartheta\left(\kappa_1 - \kappa_2\right)\sin\varepsilon. \tag{8}$$

In his first series of experiments, Eötvös has used the simple Cavendish-Coulomb balance (Fig. 4) and has performed two series of measurements choosing the East–West direction for the beam orientation: $\vartheta = \frac{\pi}{2}, \frac{3\pi}{2}$ (see Fig. 5). He puts first the same (platina) sample on both sides of the balance. In the second ("primed") series, one of them has been replaced by a different (comparison) material. From the four equations, one can express the difference $\kappa_{Pt} - \kappa_{\mathrm{Comparison}}$ through the following simple formula

$$\tau(\varphi_{\pi/2} - \varphi_{3\pi/2}) - \tau(\varphi'_{\pi/2} - \varphi'_{3\pi/2}) = 2\tau\left(\kappa_{\mathrm{Pt}} - \kappa_{\mathrm{Comparison}}\right) \tag{9}$$

Fig. 4 Left: The Cavendish-Coulomb balance of Eötvös used in the 1888–90 series of experiments. Eötvös has improved its stability through careful selection of the torsion fiber and by thermal and aerial isolation. Right: The first (asymmetric) Eötvös balance (1891) used in field experiments and gained importance in exploring oil resources in the US, India, Venezuela, etc. between 1920 and 1940. The balance was placed in the tent and observed with a theodolite

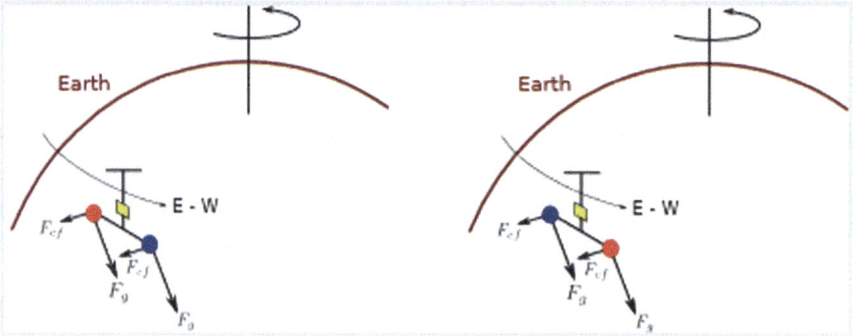

Fig. 5 Sketch of the beam orientations of the Cavendish-Coulomb balance in checking universal proportionality of inertial and gravitational masses

New versions of the Eötvös-balance have allowed between 1906 and 1909 to improve the degree of the agreement of the inertial and gravitational masses further to 1:200.000.000.

The confirmation of the equivalence principle has been acknowledged with satisfaction by Einstein in a Letter to Eötvös in January of 1918: "I cannot close my letter without expressing my gratitude for your work advancing greatly our knowledge on the identity of the gravitational and inertial masses."

Alternative gravitational theories have reinforced the significance of the Eötvös experiment. Robert Dicke and collaborators have performed in 1963 an Eötvös-type experiment in Princeton employing more advanced technology (Roll et al. 1964). They compared the gravitational acceleration of various samples attracted by the sun and improved the bound set by Eötvös by three orders of magnitude. Braginsky and Panov have performed their measurements in Moscow along the same concept in (Braginsky and Panov 1972). It is interesting that in his famous (post-hume) paper dating 1922, Eötvös and his collaborators have also considered an experiment exploiting the gravitational attraction of the sun. They recognized the great advantage of this setup, namely that the rotation of the Earth automatically exchanges the position of the samples relative to the sun in every 24 h. However, they have estimated lower sensitivity to this method compared to that in which the Earth's attraction is exploited, and the samples are exchanged by hand.

In 1986, Ephraim Fischbach has reanalyzed the original data of Eötvös and has discovered a non-trivial relation between the measured differences in the mass ratios and in the number of nucleons in 1 mol of the samples (Fischbach 1986, Fischbach and Talmadge 1999). For the interpretation of this relation, he has proposed the existence of an unknown new fundamental interaction ("Fifth Force") with a range of about 10–100 m. This conjecture has been checked by a research group led by Eric Adelberger at University of Seattle, again relying on an Eötvös-type balance (Fig. 6). During the years, this group has improved its torsion balance rotated with

Fig. 6 Eötvös-type balance of the Eöt-Wash group (University of Washington, Seattle) rotates with 1 MHz angular velocity. Its rotation period has been chosen to minimize the expected noise from the environmental oscillations and the instrument itself. The geographical location of the instrument has allowed to check the existence of any "fifth force" with longer than 1 m range of action

1 MHz angular velocity, and by 2008, they could improve upon the original Eötvös results by five orders of magnitude (Wagner et al. 2012). However, no sign could be detected for the new hypothetical fundamental force.

The search for such a new fundamental interaction hidden in the Newton-Einstein gravitational force is still in the forefront of actual researches. The latest tests were performed in space in 2017, when a new experimental arrangement, different from the torsion balance, has been employed (Touboul et al. 2019). Nevertheless, these series of experiments has been started by Roland Eötvös about 130 years ago reserving for him a rightly deserved place in the high precision exploration of fundamental forces.

2.3 Weight of a Body in Non-inertial Motion Relative to the Earth Rotation (Eötvös 1919)

Between 1901 and 1905, Oskar Hecker (Potsdam Observatory, Germany) has studied the geographical variation of the gravitational acceleration on ships moving on the sea by comparing the height of the mercury column of a barometer with the atmospheric pressure obtained from the boiling point of water. In Hecker's data, Eötvös has discovered the effect of the vertical component of the Coriolis force (Fig. 7). Hecker has reconsidered the data as suggested by Eötvös. He has evaluated also new data obtained by himself on two ships moving on the Black Sea in East–West, but opposite directions, and found perfectly consistent results for the gravitational acceleration. The modification of the effective gravitational acceleration due to the vertical component of the Coriolis-force is called *Eötvös effect*.

Eötvös made use of the phenomenon of resonance for an indoor demonstration of the Eötvös effect. He constructed a seesaw on a rotating platform placing equal weights on the two ends of the beam (see Fig. 8, left). The balance in the non-rotating state is in equilibrium, namely the two arms are in horizontal direction. When the platform starts to rotate one of the weights goes eastwards, while the other goes to

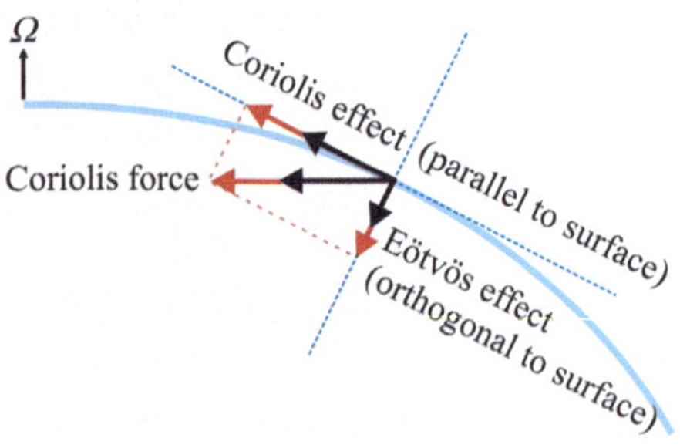

Fig. 7 Coriolis force for a body of mass m moving with velocity v eastwards at latitude φ (orthogonal to the plane of the graphics) has a component parallel to the surface deviating the motion of the body northwards. Eötvös called attention to the vertical component modifying the weight by $\Delta G = -2m\Omega \cos \vartheta\, v_{\mathrm{East}}$, where Ω is the angular velocity of the Earth rotation

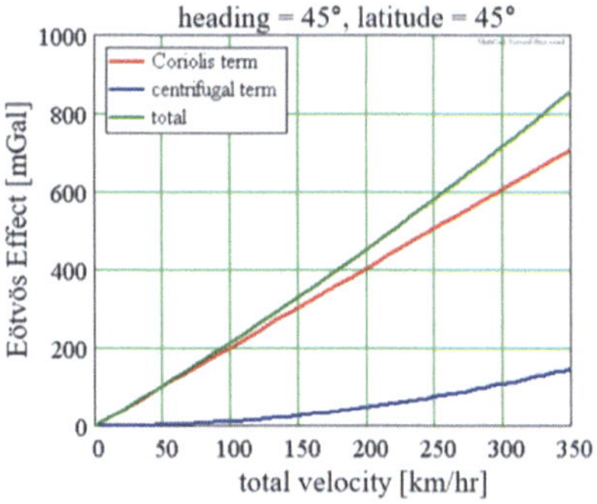

Fig. 8 Photo of the original rotating balance of Eötvös (left). Dependence of the centrifugal and the Coriolis force on the velocity of the body on the Earth surface (right)

west. In view of the Eötvös effect, the weight of the two bodies changes in an opposite way. This is a very small effect. However, if the variation of the force acting on the ends follows the eigenfrequency of the seesaw, then the amplitude of the seesaw will increase abruptly.

The rotating balance of Eötvös represents an independent evidence for the Earth' rotation, equivalent to the famous pendulum experiment of Foucault. The discussion of the Eötvös correction (Fig. 8, right) of airborne gravimetric measurements is found in all modern textbooks of geodesy.

3 Initiatives for the Modernization of Teacher's Training

The completion of the works initiated by his father in the cultural and educational policy was a central ambition of Roland Eötvös during his short ministerial position. He has achieved the parliamentary approval of the last act of religious freedom, giving equal rights to the Jewish community with other religious groups. He has extended the network of elementary education by securing the finances for 400 new school buildings. He also understood the difficult financial situation of teachers and arranged in the state budget six times higher financial support for a foundation supporting teacher's families.

Here, I wish to elaborate on his two most important initiatives which both exert important influence even today on the professional activity of the Hungarian teacher's community.

3.1 *Foundation of the Mathematical and Physical Society (1891)*

After his nomination as ordinary professor of experimental physics (1878), he has participated very actively in the discussion on the modernization of public education and of higher education. In a dispute on the unification of the different forms of secondary education, he has made the following statement: "The school needs a good school system and good teachers. You might ask which is more important. Myself, I would prefer the latter." He has added that in a suboptimal system the good teacher finds the way to teach well, but without good teachers even the best system will fail. Another time he has expressed his credo as it follows: "The quality of teaching depends in first place on the scientific preparedness of the teachers."

This sentence is a compressed expression of his "teacher-scientist" concept. According to Eötvös, teachers should be able to follow the frontlines of science. He has proposed the foundation of the Mathematical and Physical Society as a medium where this sort of life-long self-education can be realized. Upon the foundation of the society, also its monthly periodical has been launched (Fig. 9). The present journal of the Roland Eötvös Physical Society, "Physics Review," has a separate section on Physics Teaching with its own editors and publishes every month 2–3 papers, mainly by practicing teachers and PhD-students developing innovative physics teaching methods and equipments for the demonstration of exciting natural phenomena.

3.2 *Foundation of the József Eötvös College (1895)*

This teacher's training college has been founded as an independent institution in form of a boarding school (Fig. 10). Its 100 members came from the countryside. 30 places were given to students from poor families without tuition fee payment obligation. The selection system of the new college members and the very strict requirements concerning the number of subjects to be studied and the performance level at the exams followed largely the norms of École Normale Supérieure (Paris). The college had from the very start one the best scientific libraries of the country, and the students could attend language courses by lectors invited from countries where the given language is (one of) the official language(s).

Among the pupils of the nearly six decades during which the college has followed closely the original prescriptions of Eötvös, one can find the worldwide known composer Zoltán Kodály, but also the internationally esteemed mathematics teacher Tamás Varga or the legendary physics teacher Miklós Vermes who was the head of the jury of the Eötvös Physics Competition (1894) for half century.

The role of teachers is heavily disputed nowadays. Followers of some fashionable pedagogical directions consider the teacher-scientist concept of Eötvös a kind of nineteenth-century elitist ideal. But this ideal is still attractive: after the collapse of

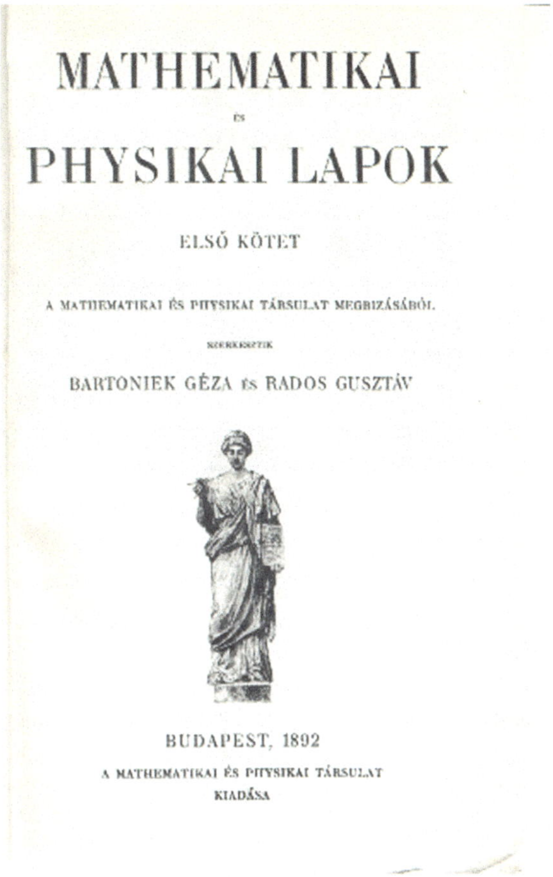
MATHEMATIKAI
ÉS
PHYSIKAI LAPOK

ELSŐ KÖTET

A MATHEMATIKAI ÉS PHYSIKAI TÁRSULAT MEGBIZÁSÁBÓL.

SZERKESZTIK

BARTONIEK GÉZA ÉS RADOS GUSZTÁV

BUDAPEST, 1892

A MATHEMATIKAI ÉS PHYSIKAI TÁRSULAT
KIADÁSA

Fig. 9 First issue of the mathematical and physical journal, where Eötvös has stated: "We shall write this journal for ourselves, not to publish original new results, but for presenting the latest results of the world science pedagogically, that every teacher could follow and make use of it in the teaching procedure"

the "socialist" system which based its ideology on egalitarian pedagogical principles, new colleges were founded having in mind the example of the József Eötvös College. One of the first was the Bolyai College of the Faculty of Science of Eötvös University started in 1992.

A particularly important challenge concerns the way one should understand today the concept of "teacher-scientist." An initiative has been taken by the Hungarian Academy of Sciences in 2016, a program of Content Pedagogy development research. The goal is to propose ways to resolve hot problems of science teaching. Only project propositions with considerable participation of active high-school teachers could have chance to receive the 4 years support. Results obtained by the members of the Physics Education Research Group (one of the 19 supported research groups) have been presented in sections of this conference. The presenters can be considered teacher-scientists of our days.

In summary of this part, I wish to emphasize that the activities, the institutions and even the disputes demonstrate that the pedagogical ideas of Eötvös are still alive.

Fig. 10 Building of the Eötvös College (1910)

3.3 Conclusion

This short lecture could not deal with all aspects of the colorful life of Roland Eötvös. There was no time and space to describe neither his science popularization activity nor his impressive alpine mountaineering record, documented through his stereographic photos made during his summer holidays in the Dolomites. I have put emphasis on his scientific achievements and on his initiatives improving the education and professional organization of physics teachers in Hungary. Telling his science story to our students should attract them toward the advances of modern gravitational physics. Introducing his character will positively impact on the personality of the young generation of teachers and researchers.

References

Braginsky VB, Panov VI (1972) Soviet Phys JEPT 34:463

Dobszay T, Estók J, Gyáni G, Patkós A (eds.) Roland Eötvös Memorial Album. ISBN 978-963-09-9929-8. Kossuth Publishing House, Budapest 2019, freely downloadable from http://real-eod.mtak.hu/3799/1/001-176_tordelt1_ANGOL.pdf

Eötvös RV (1886) Annalen.d. Physik 263:448

Eötvös RV (1919) Annalen d. Physik 364:743

Eötvös RV, Pekár D, Fekete E (1922) Annalen d. Physik 373:11–66

Fischbach E, Talmadge CL (1999) The Search for non-Newtonian gravity. Springer, Heidelberg, New York

Fischbach E, Sudarsky D, Szafer A, Talmadge CL, Aronson SH (1986) Phys Rev Lett 56:1427

Fröhlich I (ed.) (1930) Báró Eötvös Loránd Emlékkönyv (in Hungarian), MTA, Budapest
Roll PG, Krotkov R, Dicke RH (1964) Ann Phys 26:442
Touboul P et al (2019) Class Quantum Gravity 36:225006
Wagner TA, Sclamminger S, Grundlach JS, Adelberger EG (2012) Class Quantum Gravity 29:184002

The Impact of the Equivalence Principle on Physics Teaching—The Ongoing Opposition in Teaching of Weight-Gravity

Igal Galili

Abstract It is common to praise big historical events in physics especially on the occasion of their centennial. It is not common, however, that the contribution to modern physics is relevant to the basic physics curriculum of any school. This happened to be the case of principle of equivalence and its exact confirmation provided by Roland von Eötvös a century ago (1909), which projected to the cardinal change in the concept of weight as different from the gravitational force. This essay briefly depicts the history of the concept of weight-gravity and its impact on physics teaching in regular school classes. The controversy of gravity versus gravitation introduces to the learners the major feature of physics nature—its commitment to the operational definition of physics concepts. What makes the history of weight especially interesting is that the debate on weight definition in physics teaching is not closed and, in fact, splits the community of physics educators around the globe into two groups: Newtonians and Einsteinians. This makes our life interesting and us alive. We suggest a solution.

Keywords Weight · Gravity · Gravitation · Concept definition

1 Introduction

In the wide range of physics curricula, it is upon the teacher to prove that all bodies, regardless their mass, fall with the same acceleration when dropped from the height. Physics teachers write the second law of motion with the force of gravitation, both bear the name of Newton, and cancel mass on both sides of the equation:

$$ma = G\frac{mM}{r_{1,2}^2},\ a = G\frac{M}{r_{1,2}^2} \tag{1}$$

I. Galili (✉)
The Hebrew University of Jerusalem, Jerusalem, Israel
e-mail: igal.galili@mail.huji.ac.il

B. Jarosievitz and C. Sükösd (eds.), *Teaching-Learning Contemporary Physics*, Challenges in Physics Education,
https://doi.org/10.1007/978-3-030-78720-2_7

Wonderful! It takes a minute to get the amazing result: any object, regardless its mass falls with the same acceleration. What a surprise! A question arises, however, were they all so stupid not to *see* it for two thousand years? It is so simple… Nothing can be worse than this impression of the student. Therefore, a good teacher stops there and asks, is it all right? On the left, we have inertial mass—m_i—responsible for inertia; while on the right side, there is the gravitational mass—m_g—responsible for the attraction.

$$m_i a = G\frac{m_g M_g}{r_{1,2}^2} \tag{2}$$

Who said they are the same? This question puzzled Newton, and he knew where to look for the answer—experiment. Newton was among the very few people who were extremely good in both theory and experimentation. He tried to check the relation between the inertial and gravitational masses with the tools at hand (simple pendulum) and revealed equality. He was amazed, but had nothing to add to his surprise, this is it… God's design… The claim of this identity (Eq. 3) became fundamental and remained without much attention of physicists until twentieth century.

$$m_i = m_g \tag{3}$$

There is, however, another form of the same equivalence which is unfortunately not common at schools. This is the perspective of different frames of reference, considering the accounts of reality by different observers. The first to suggest it was Huygens, a renowned contemporary rival of Newton.

Huygens introduced the rotating frame of reference and a new force which manifested itself in such account—centrifugal force. Later, it was labeled as inertial force. He proved, basing on Galileo's kinematic law of free falling, that centrifugal force is identical to gravitational. The equivalence $m_i = m_g$ got the new form:

$$\boldsymbol{F}_i = \boldsymbol{F}_g \tag{4}$$

It is in this form of Eq. 4; however, the principle of equivalence became the claim of the new theory, the theory of General Relativity of Einstein in 1916, addressing the account of reality in the frame of reference under acceleration (Einstein and Infeld 1938). In that theory, in order to reproduce the principle of equivalence, Einstein related directly the factor determining the motion of a body, m_i, in the curved space–time, with the factor determining that curvature—m_g. If the theory is valid, which is a matter of experimental verification, then, the principle of equivalence indeed holds, and vice versa.

Yet, the path of history was different. By the end of the nineteenth century, Roland von Eötvös in Budapest posed the question of empirical verification of the equivalence of inertial and gravitational masses with the tools available about two hundred years after Newton. He succeeded to improve the numerical accuracy of the claim to so high level that for the contemporary physics it was sufficient to consider the equivalence of the masses. It was this fact that was taken by Einstein as given. Then, the theory was produced, and its corroboration in several famous experiments implied the truth of the equivalence principle placed in the nucleus of this fundamental theory of physics (Born 1962).

In physics, the forms of Eqs. 3 and 4 are equivalent. Yet, it is not so in schools physics. It is common that school curricula deal mainly with classical Newtonian physics and ignore non-inertial frames of reference and so the inertial forces. Indeed, these forces, centrifugal, and Coriolis are non-Newtonian, that is, they have no partners of interaction. Labeled as fictitious or pseudo-forces (not real!), they were left out of school teaching seeking the maximal simplicity and univocal accounts. In reality, however, this strategy implied the demand of the proper choice of an observer: inertial—yes, non-inertial—no. This requirement of proper selection appeared to be non-trivial for many students, causing much of confusion (Galili and Kaplan 2002). In fact, we are never in the situation without any acceleration, and the frame of reference related to the Earth ground is not inertial too, as evidenced by various observed phenomena, weight change with geographical location, gigantic whirls in atmosphere, Coriolis effects on moving objects, and water streams. In effect, the account of situation using inertial forces is often much easier and closer to intuition of the students. Inertial forces are used by any engineer or technician, pilot or navigator, driving instructor, or sailor. In short, it is used everywhere, but not in high school physics class even with advanced curriculum. As a result, naïve intuition surpasses in modernity the knowledge of school graduators. Inertial forces can bring students closer to modern physics not only through learning about the equivalence principle but by putting to the fore the operational definition of force and familiarizing with the status of physics concepts as observer dependent.

Whether or not one includes inertial forces and the principle of equivalence into physics teaching at schools, there is still a concept which went through a radical change following the new epistemology. This is the concept of weight. Perhaps not too important in some domains of advanced physics, its inextricable presence in everyday life prevents any physics curriculum to ignore it. In fact, one can use its teaching as an introduction to the new approach to concepts in modern physics. It happens that it presents a special problem of contemporary physics teaching.

2 Brief History of the Weight Concept

Since Aristotle, we know that in order to get understanding of a subject matter, we need to follow its history. In the most concise form, the history of weight in physics exhibits three big periods (Galili 2001).

The *first* period started in the physics of Classical Greece and lasted to the seventeenth century. Although there were interesting variations in understanding the nature of weight, one may *here* roughly address this knowledge as ascribing weight to the intrinsic feature of matter we all perceive—its heaviness or gravity. One may mention the four Aristotelian elements of heavy earth, less heavy water, light air, and superlight fire in Hellenic science. Though weight caused natural falling, it was not a force. The conception was interesting: "what is natural does not require being forced" (Aquinas 1267).

In the following Hellenistic science, the understanding progressed to the operational definition of weight as the heaviness of an object being measured in its weighing (Euclid and Archimedes) (Clagett (ed.) 1957). The notion of density refined the holistic understanding of weight removing the confusion with size. Weight concept was further promoted, mainly in medieval science, by ascribing additional (*accidental*) weight to a moving (falling) body. Such weight was often related to impetus, on top of the *natural* weight. This or other way weight addressed the virtue of *heaviness*. In different contexts, the terms weight, gravity, ponder, and mass were in common use creating a family of connotative notions with close meaning.

The *second* period started with the scientific revolution of the seventeenth century. Galileo still expressed a tactile perception of weight as diminished when an object is immersed in liquid, imagine the effort in supporting objects in water. Yet, the gross change in weight understanding was due to Newton (Newton 1687). As we teach in classes since then, Newton established a new fundamental theory of the world. Within the new order, he introduced gravitational force as an agent of universal attraction between any two objects (Eq. 5a). In the context of our review, we should emphasize two aspects of his great accomplishment. Firstly, he separated between inertial mass and weight of any object. Secondly, he explained weight (heaviness) by equating it to the gravitational force exerted on the object – a great invention indeed (Eq. 5b):

$$F_g = G\frac{mM}{r_{1,2}^2} \tag{5a}$$

$$\Rightarrow \boldsymbol{W} = \boldsymbol{F}_g \tag{5b}$$

It is important to see that the Newtonian understanding was introduced while building a theory of the solar system, the account of planetary motion as depicted by Kepler's kinematic laws. It was a grand picture observed by the unique observer, God, from above of his creature—the Universe. No other observers were even considered. In this basic context, the result of weighing of an object was secondary to the major theoretical claim of the grand law. Weighing as a local measurement was explained by the theory (not vice versa!). Newton addressed the confusion of Galileo regarding the weight of the objects immersed in water: the *true* weight remained unchanged and equal to the gravitational force. It is the *apparent* ("common", illusionary, perceived) weight that changed. Moreover, Newtonian weight was introduced in pairs of equal forces exerted on a couple of interacting objects, "weight towards..." Therefore, the weight of the Earth toward the Sun was different from the weight of the Earth toward the Moon. The weight of an apple was equal to the weight of the Earth. There was no true weightlessness in this universe, but perhaps approximate one "far away" from any object.

The *third* period of weight understanding emerged in the twentieth century with the new grandiose scientific revolution. With the introduction of the *General Theory* of *Relativity*, the principle of equivalence between inertial and gravitational masses came to the fore. At the same time, a parallel revolution took place in physics epistemology—the positivistic requirement of the operational definition of physical concepts. It was launched by Ernst Mach and masterfully applied by Albert Einstein in his revolutionary *Theory of Special Relativity* in 1905. In it, Einstein operationally defined simultaneity. Drawing on two postulates as foundation, he elicited a whole new physics of space–time, an interconnected collection of world pictures by different observers (Born 1962).

Hans Reichenbach while considering the philosophy of the Einsteinian revolution in relativity applied the operational approach to the concept of weight (Reichenbach 1928). Unlike the pantheist approach of Newton, Einstein was far more modest. Physics theory now had to be fully constructed in a closed laboratory while using solely local measurements. With regard to weight, one had to define the concept solely based on the appropriate measurement—weighing. This approach posed a question regarding the previous weight definition of whether gravitational force provides a unique contribution to the result of weighing. In light of the equivalence principle, the answer was negative; especially that one can be never sure regarding the status of his/her closed laboratory with respect to its acceleration. In the new vision, weight is caused by both gravitation and inertial forces in an unknown combination. In the state of rest in a laboratory, the weight is equal to the sum:

$$\boldsymbol{W} = \boldsymbol{F}_g + \boldsymbol{F}_i \tag{6}$$

In the striking difference to the stage two, in the stage three of weight account, the state of weightlessness became possible in no relation to distance, but due to the nullifying of the force exerted on the support of an object, directly measured. This happens in a free gravitational motion, or falling, in the common use of this word. In other words, weightlessness corresponds to the situation of zero result in a standard weighing.[1]

The new account heralded a historical divorce of weight from gravitational force, after their marriage in the seventeenth century. We may thus summarize the history of weight concept in the diagram of Fig. 1 displaying three big periods in understanding of weight concept. In the first period, weight was identified with a specific characteristic of an object that depended on the nature of its material. In the seventeenth century, following the universal account of reality by an inclusive theory—the account of classical mechanics, weight was split from the concept of inertial mass (scalar) and identified with the gravitational force exerted on an object. The second period lasted to the following revolution in science, in the twentieth century.

[1] The requirement of *standard* excludes any kind of additional support such as, for instance, in weighing of an object immersed in water.

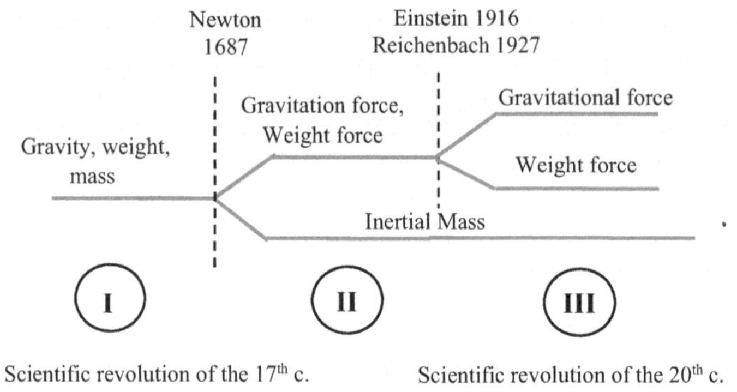

Fig. 1 Periods of conceptual evolution of the concept weight through the history of physics

3 The Weight Concept in Physics Education

Thus, due to physics philosophy, the transfer from the stage two of nominal definition of weight was quick, from 1916 to 1927. That was in physics. Yet, the transfer from physics to physics education became tremendously slow and actually is undergoing in our days. One may explain that slowness by natural mainly social factors. Education has a huge mass for involving a great number of curricular designers in numerous countries around the globe. Physics educators naturally preserved the highly entrenched in many years of use gravitational definition. Global modifications of the way of teaching are problematic.

To our knowledge, the first to adopt the operational definition of weight in regular school teaching was Baruch and Vizansky in 1937 in Israel (Baruch and Vizansky 1937). This innovation was, however, suppressed by the overwhelming influence of the famous Sears and Zemansky Physics Course from the USA, which kept with the gravitational definition of weight (Fig. 2a). The breakthrough was apparently reached by Chaikin in the USSR in 1947 (Chaikin 1947). Due to the centralized system of education, millions of students were exposed to the new curricular content—the operational definition of weight (Fig. 2b) as soon as the academy decided so. In the USA, the call for teaching operational definition of weight appeared from 1962, and the first author adopting it was seemingly Orear in 1967 (King 1962). The definition received legitimacy from the respected university professors in Maryland, Marion, and Hornyack (Marion and Hornyack 1982). French and Levin in MIT introduced a modified operational definition in which weight was identified with the force directly perceived by a person standing on the ground (French 1971). Later, the operational definition became more common in and spread among introductory courses of physics including Europe (Lerner 1996). Though slowly, the transition to the operational definition of weight is undergoing. One may observe it in the popular physics textbooks by Halliday and Resnick, Hewitt, and Knight who all in their earlier editions employed the gravitational definition of weight (Resnick and Halliday 1988). They later switched to the operational one (Halliday et al. 2001). Figure 3 compares

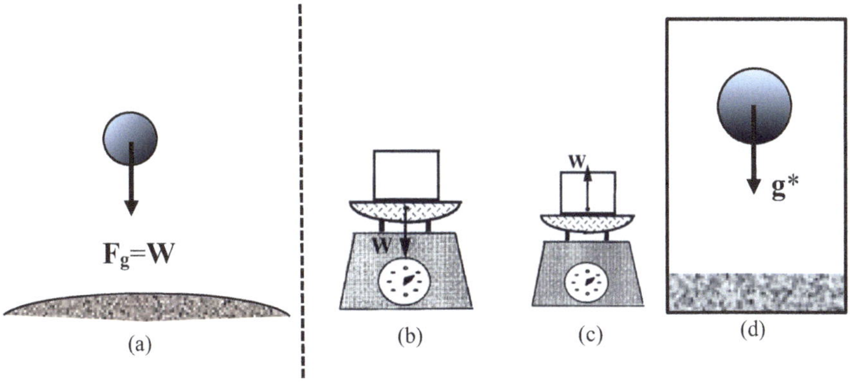

Fig. 2 Weight definitions in physics textbooks: **a** weight **W** defined as the gravitational force on an object exerted by the Earth; **b** operational definition: weight **W** defined as the force on a calibrated scale. **c** Operational definition: weight **W** defined as the force on the object by the support. **d** Operational definition: weight defined as the force causing free falling in a certain closed laboratory (the local acceleration g^*)

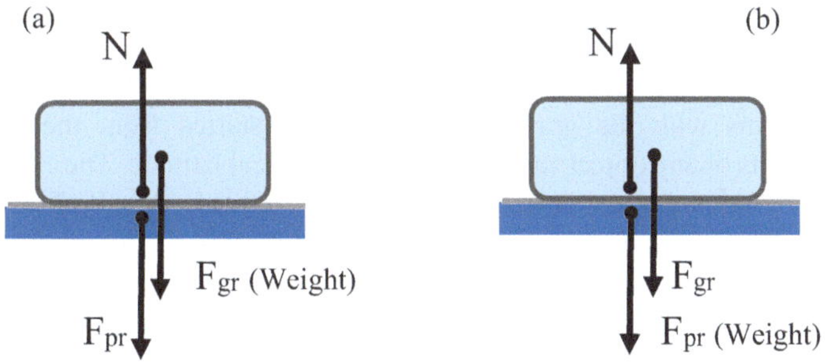

Fig. 3 Comparison between weight **W** definitions **a** The case of the gravitational definition. **b** The case of operational definition, N—normal force, F_{gr}—gravitational force, F_{pr}—pressing force

the identification of forces as used in the gravitational and operational definitions of weight.

It is of special interest to mention the textbook by Keller et al. (1993) who in 1993 employed still another modified operational definition of weight. In it, weight was determined by the effective force exerted on an object causing a free falling in a closed laboratory (Fig. 2d). The authors reminded to all their readers that in fact, *many years ago*, the general conference on weights and measure (CGPM) adopted as International Standard in Mechanics (ISO 80000-4) the operational, not gravitational definition of weight (Fig. 2d) (Taylor and Thompson 2008). It prescribed it to all practical goals which might be illustrated by geodesic search for underground ores indicated by period changes of a regular pendulum located above a specific location. This definition, however, ascribed weight of an object to the net gravi-inertial force on it. It can be shown that all three operational definitions, b, c, and d, provide equivalent magnitudes of the measured weight.

Yet, the great most of physics textbooks in many countries currently employ the gravitational definition of weight. It, however, was changed from the form introduced by Newton. Facing the apparent mismatch between the weighing results and the gravitational force calculated according to the universal law of gravitation, the concept of weight was split to *apparent* weight (the result of weighing) and *true* weight (the net gravitational force acting on the object) (Young and Freedman 2012). While the apparent weight coincided with the weight operationally defined (often unclear regarding the body that the weight force is applied to…), the problem of how to measure the true weight remained, as it was before regarding weight.

To appreciate the problem of the gravitational definition, let us quote the definition of weight as it continues to appear for many years in the textbooks of Sears and Zemansky and currently appears in the text by Young and Freedman (2012):

> The weight of a body is the total gravitational force exerted on the body by all other bodies in the universe

This definition though could have some meaning in the rationalist philosophy in the past, has no meaning in physics of today. Indeed, the weight defined this way cannot be calculated, estimated, measured and has no use in any physics problem. Such a definition simply does not belong to physics as we understand it now.

The problems with the gravitational definition started from the beginning. Weighing results of any object vary with the geographical latitude. The change of the free fall acceleration g is $\Delta g = 0.043\ \mathrm{m/s^2}$ ($\Delta g/g = 0.0044$). A half of this effect is due to the Earth rotation. Could it be neglected? Not in trade… (A ton of cargo from Stockholm "loses" 4 kg after arriving to Somalia.) Tables of the geographic variation of (effective) g were traditionally used for many years; weights were marked according to place of their production, and people did not care regarding the origin for this inconvenience. Newton's theory of gravitation was irrelevant for practical activities where *effective g* replaced the theoretically calculated *true g* (French 1983).

4 The State of Weightlessness

However, the major conceptual difficulty came from considering the state of weightlessness. It presents a special problem to account for the situation in which an object is in a free fall or its motion is solely under the influence of gravitation. Such is the motion of a planet or its satellite. Such is, for example, the motion of the international space station orbiting the Earth for years already. In this case, the difference between the two approaches to weight becomes striking.

While this state is considered to be the state of a *true* weightlessness if weight is identified operationally, in the gravitational account, it is only an *apparent* weightlessness—"weightlessness"—a mere illusion of lacking weight (Fig. 4). It is only the apparent weight of the objects inside the satellite that nullifies, while their true

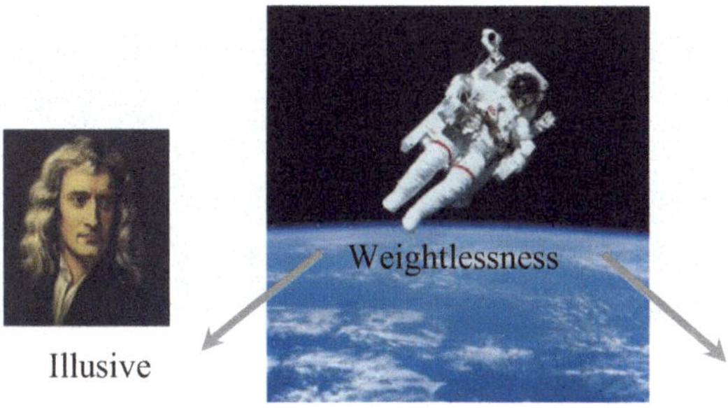

Fig. 4 Two accounts of the state in which an object is in a free gravitational motion—weightlessness (Picture credit: NASA)

weight (the gravitational force) could be almost as it was on the surface of the Earth. Clearly, this gross mismatch brought a permanent challenge to teachers to explain and to students to understand (Galili 1995). It is with regard to weightlessness that the split in the world of physics education, to Newtonians and Einsteinians, came to the fore.

The gravitational definition of weight can provoke misunderstanding of weight and gravitation (Galili et al. 1996). Consider the situation with zero net force of gravitation. This case was famously depicted by Joule Verne, a good student of Newton, who in 1865 sent his heroes to reach the Moon inside a big shell as a spacecraft. Verne depicted the state of weightlessness as if taking place at the point where the attraction to the Earth and to the Moon canceled each other (Fig. 5a). There, in that very point only, Verne described a very strange experience of the astronauts floating in the cabin being weightless (Fig. 5b). It lasted only for an instant after which the astronauts gradually restored their weights, this time, mainly toward the Moon. As we all can observe on our days, nothing could be more distant from reality. Staying on the ballistic trajectory toward the Moon (or anywhere else), the astronauts did not even know that they passed this point being weightless all the time.

The state of weightlessness is more than just a topic in the contemporary world of space exploration. On our days, it is the everyday reality of continuous life of astronauts in the international space station observed in news. Ignoring this topic by numerous physics textbooks and educational programs looks, to say the least, strange and inappropriate (Rutherford 1990). Evidently, weightlessness—the strange life without weight—attracts curiosity of all students and not only students. A whole new science of "microgravity," as it termed in the materials published by National Aeronautics and Space Administration (NASA), explores the vast variety of phenomena in the world without gravity from the perspectives of physics, chemistry, biology, material science, etc.

A special importance of this knowledge emerges when one considers the impact of weightlessness on the human organism (Thornton and Bonato 2017). It appears that beyond the apparent fun of floating, there is a major threaten. A long stay in weightlessness causes serious damage of decalcination and mass loss of bones, disorder

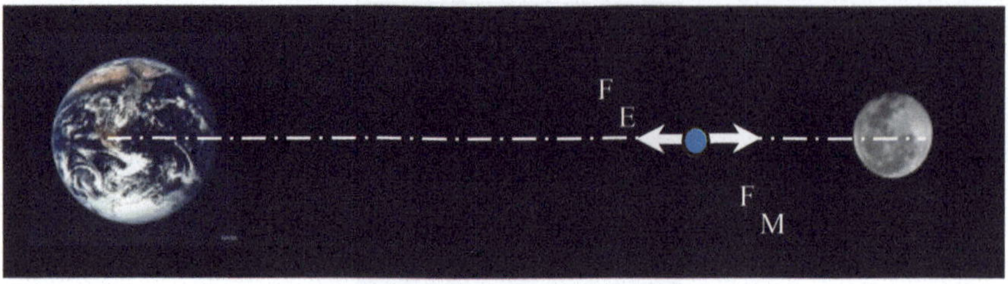

(a)

(b)

Fig. 5 **a** Point between the earth and the moon where the attractions to the earth and to the moon cancel each other. **b** The state of weightlessness at that point lasting for an instant as was described by Jules Verne in 1865

of vestibular functioning and vision, muscular atrophy, blood vessels contraction, the change in chemical reactions implying changes in metabolism, interaction with microbes, and a strong psycho-physical impact. The time scale of significant impact counts months. In accordance, the stay of researchers in the station is limited. The educational inference to make is that biological life requires weight and not gravitation!

Humans are *homo gravitas*—they essentially need weight, and it can be created without gravitation, but due to rotation. One may provide a clue of non-gravitational weight in a demonstration induced by Newton. He considered the impact of central force causing deflection of a moving object due to frequent strikes which deflect it from straight motion. In the limit of very frequent strikes, it is similar to continuous pressure of the wall on a ball (Fig. 6a). Newton would call the force on the *wall centrifugal* force. Yet, it fits to the operationally defined weight. In the twentieth

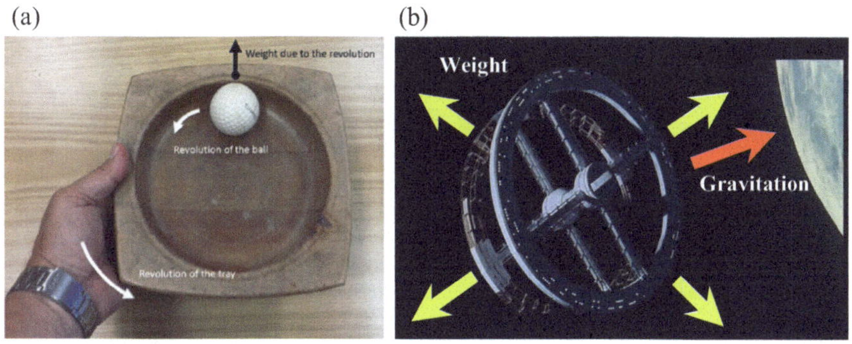

Fig. 6 **a** Demonstration of the weight due to revolving of the object used in the instruction to explain weight in the rotating space station. The ball presses the curb of the revolving tray with weight force. **b** Weight (yellow arrows) and gravitation (red arrow) in the environment of rotating space station (The background image is a screen shot from NASA website)

century, this idea produced a technological solution for creation weight in space—a rotational station. One cannot imagine a situation better illustrating the difference between weight/gravity and gravitation (Fig. 6b). The account does not require anything beyond elementary algebra. It provides weight of a body:

$$\boldsymbol{W} = m\omega^2 \boldsymbol{R} \qquad (7)$$

5 New Teaching Approach

Importantly, using inertial force is not a condition to explain the operationally defined weight. One may continue the teaching keeping with Newtonian framework, while defining weight as the force exerted on the support, rather than to the gravitational force. To bridge to the common situation, one may slightly modify the speech: "weight of a bodies is often *due* to gravitation" instead of "weight *is* the gravitational force." This change prevents the logical fallacy of equating a cause with its effect.

The replacement of gravitation with "artificial" gravity will be used in the very distant future of space colonies where people will live far away from the Earth. In the much close future, it will happen in the long interplanetary voyagers, to Mars and beyond.

It might look that the pedagogical advantages as well disciplinary correctness of the operational over gravitational approach to the weight concept are rather obvious. Again, it is independent of whether physics teaching adopts inertial forces or not. Yet, the optimal teaching of weight is, in our belief, different in another dimension. We argue for the discipline-cultural approach to physics curriculum (Galili and

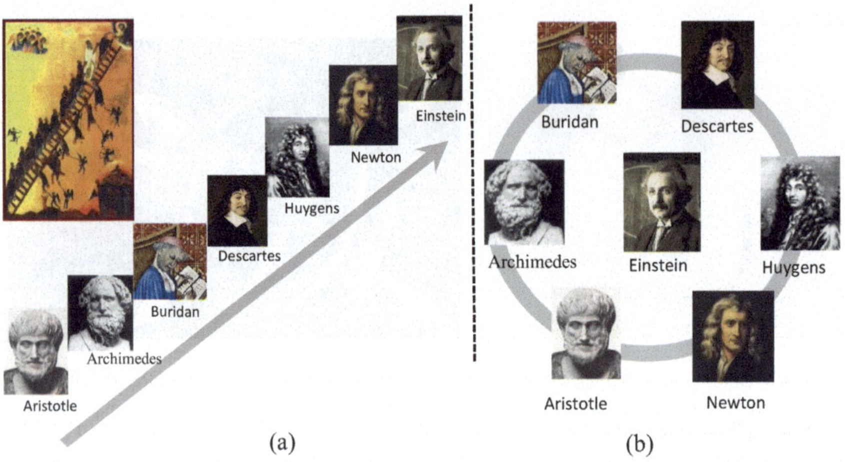

Fig. 7 **a** Ladder type of addressing steps of development of the knowledge regarding gravity. (The Orthodox church icon represents the idea of ladder climbing in theology. The parallel includes the failure on the way up.) **b** Presenting the dialogical nature of the scientific knowledge of gravity within the paradigm of discipline–culture. It reveals a diachronic dialogue of conceptions which constitutes physics knowledge as a discipline-culture

Matthews 2017). From its perspective, the teacher presents the conceptual evolution of the understanding gravity, which displays more than the goal conception. It is rather obvious, especially, with regard to weight, that teaching should not ignore the major ideas in the consolidation of the present understanding. It has been understood that learning should consider the way certain knowledge was constructed rather than imposed. The discipline-cultural teaching reveals dialogical nature of the subject matter which is more than informing about the historical steps in the knowledge growth. Rather than a ladder of climbing while eliminating the previous steps as fallacies (Fig. 7a), the proper teaching revives a diachronic dialog, the debate revealing the rationale. It is this debate of conceptions that constitutes the nature of science. Besides being historically authentic, it happens that this very approach essentially promotes the meaningful genuine learning too also because it resonates with the intuitive conceptions of the students, and this way promotes their conceptual change.

In its minimalistic form, the discipline-cultural curriculum would include the ideas regarding gravity produced in several scientific periods, employing different worldviews. It would include the basic conceptions developed by Aristotle, Archimedes, Buridan, Descartes, Huygens, Newton, and Einstein (Fig. 7b).

6 The Weight Concept in Physics Education Research

Physics education research regarding weight concept draws on the philosophical and pedagogical requirements of the operational definition of physics concepts (Margenau 1950). Within this understanding, the subject of weight understanding in the course of learning was explored. The following briefly presents the product of that effort.

Firstly, the pertinent students' conceptions were elicited in different populations of school students. The elicited knowledge is better represented if it is structured using two-level taxonomy of scheme-facets. In it, schemes-of-knowledge presented context independent conceptions of accounting reality. Each scheme was related to a cluster of affiliated facets of knowledge which were context dependent. The facets of the cluster drew on the claim of the correspondent scheme as if being its concrete manifestations. Scheme and facets could be true or false with respect to their scientific correctness.

It appeared that the younger students of elementary school (Piaget 1972) prevailingly held such schemes of weight as muscular effort to support or move objects, pressing force and heaviness of objects, and weight non-conservation. The schemes implied such facets as weightless dust, vapor, and other naturally "light objects," weight dependence on the shape and size of the object held, weight as resistance to motion and cause floating in water. All of the conceptions apparently drew on the concrete tactile experience, and none of them was related to gravitation. They considered weight as an intrinsic quality of the object, by that reminding the first historical conceptions of weight (Stage I in Fig. 1).

Students of middle and high schools (Galili 1995; Galili et al. 1996; Ruggiero et al. 1985) instructed in gravitational definition displayed quite different conceptions elicited from their accounts for a variety of physical situations presented in assessments. We may exemplify a few of students' conceptions (Table 1). The deficiency of the resulted knowledge could be related to the univocal instruction within the gravitational approach and failure to apply true and apparent weight dichotomy. The relation between weight and weighing results was apparently not discussed in classes and so remained obscure for the students.

At the next stage of investigation, we tried to assess the impact of the new teaching involving the operational definition of weight being introduced and compared with the gravitational definition. The result obtained in small groups of directed instruction employing dialogical pedagogy (Stein et al. 2015) as well as in lecturing regular classes (Galili et al. 2017) indicated impressively strong decrease in the frequency of erroneous conceptions of weight and gravitation.

In our investigation of teachers' knowledge and opinions on the subject (Galili and Lehavi 2003), we found the need for addressing concept definitions, in particular, the relationship between the kind of adopted weight definition and understanding the results of weighing. In addition, in that study, a synergetic instruction of weight, being contrasted with the tidal forces, was suggested and justified.

Table 1 Scheme-facets structure of students' knowledge about weight

	Scheme of knowledge	Facets of knowledge
1	Weight is determined by distance	Satellite (and objects inside) is (almost) weightless Objects "in space" are weightless, and they are far away from the Earth Weight increases in going below the Earth surface and becomes infinite in its center
2	Vacuum causes weightlessness	Objects are (almost) weightless on the Moon, in spaceships, in space, under glass dome with evacuated air
3	Gravitation can be neutralized by buoyant force	Swimmers in water have lower weight/are lighter Floating in water indicates weightlessness
4	Weighing informs about the gravitational force on the object	Weighing on the Earth should take into account attraction to the Moon (or the Sun)

Moreover, the teachers argued for a special importance of concept definitions in general and complained on lacking appropriate content in their training program (Galili and Lehavi 2006). This their view rather contradicted to the popular perception of the hierarchy of cognitive activities in the process of learning. The Bloom taxonomy of cognitive activities (Anderson 2013) implied very low importance to concept definitions, placing the skill of defining among low order cognitive skills, knowledge, and understanding. The presented here analysis of the weight concept, as well as the opinions of physics teachers of our sample, clearly testify that considering the weight concept definition and its applications involve analysis, evaluation and problem solving skills which were referred to by the same cognitive taxonomy as higher order cognitive skills.

We may conclude with expressing hope that the presented here multifaceted analysis of the concept of weight in physics education will inspire our colleagues to join considering this subject by relevant studies which ultimately will promote conceptual upgrading of the contemporary physics curriculum. Such progress regarding the weight concept will be in the spirit of the fundamental change which took place in physics due to the theoretical innovation by Albert Einstein and its experimental confirmation by Roland von Eötvös which hundredth anniversary we celebrated in this conference in Budapest.

References

Anderson LW (2013) A taxonomy for learning and teaching. Pearson, Harlow, UK, pp 238, 240, 272; Krathwohl DR (2002) Theory into practice, pp 414, 212–218

Aquinas T (1267/1952) Summa Theologica. Encyclopedia Britannica, Chicago, pp 541–542

Baruch A, Vizansky A (1937) Physics. Dvir, Tel-Aviv

Born M (1962) Einstein's theory of relativity. Hafner, New York
Chaikin SE (1947/1963) Mechanics. Nauka, Moscow
Einstein A, Infeld L (1938) Evolution of physics. Cambridge University Press, Cambridge, UK
French AP (1971) Newtonian mechanics. Norton, New York, p 229; Levin W (2016) MIT Phys Lect. https://www.youtube.com/watch?v=M0mxyPOMcw0
French AP (1983) Is g really the acceleration due to gravity? Phys Teacher 21:528–529
Galili I (1995) Interpretation of children understanding of weightlessness. Res Sci Educ 25(1):51–74
Galili I (2001) Weight versus gravitational force: historical and educational perspectives. Int J Sci Educ 23(10):1073–1093
Galili I (2017) Scientific knowledge as a culture—a paradigm of knowledge representation for meaningful teaching and learning science. In: Matthews MR (ed) History, philosophy and science teaching research new perspectives. Springer, Dordrecht, pp 203–233
Galili I, Kaplan D (2002) Students' interpretation of water surface orientation and inertial forces in physics curriculum. Praxis Der Naturwissenschaften Physik in Der Schule 51(7):2–11
Galili I, Lehavi Y (2003) The importance of weightlessness and tides in teaching gravitation. Am J Phys 71(11):1127–1135
Galili I, Lehavi Y (2006) Definitions of physical concepts: a study of physics teachers' knowledge and views. Int J Sci Educ 28(5):521–541
Galili I, Kaplan D (1996) Students operation with the concept of weight. Sci Educ 80(4):457–487; Galili I, Lehavi Y (2003) The importance of weightlessness and tides in teaching gravitation. Am J Phys 71(11):1127–1135
Galili I, Bar V, Brosh Y (2017) Teaching weight-gravity and gravitation in middle school—testing a new instructional approach. Sci Educ 26(3):977–1010
Halliday D, Resnick R, Walker J (2001) Fundamentals of physics. Wiley, New York; Walker J (2008) Halliday-Resnick fundamentals of physics. Wiley, New York, p 95; Hewitt PG (2013) Conceptual physics. Pearson, San Francisco (2006); Knight RD (2013) Physics for scientists and engineers. Pearson, Reading, MA, p 146
Keller FJ, Gettys WE, Skove MJ (1993) Physics. McGraw Hill, New York
King AL (1962) Weight and weightlessness. Am J Phys 30(4):387; Iona M (1975) The meaning of weight. Phys Teacher 13:263; Weightlessness is real 25(4):418 (1987); Hewitt PG (1967) Phys Teacher 51(6):330 (2013); Orear J Fundamental Phys:82 (1967)
Lerner SL (2009) Phys; Jones and Barlett, Sudbury MA (1996); Lopas V (2009) Pedagog Rev 19(2);147–164
Margenau H (1950) The nature of physical reality. McGraw-Hill, New York, pp 220–244; Karplus R (2003) Introductory physics: a model approach. Benjamin, New York; Arons A (1990) A guide to introductory physics teaching. Willey, New York
Marion JB, Hornyack WF (1982) Physics for science and engineering. Saunders, New York, p 129
Newton I (1687/1999) Mathematical principles of natural philosophy. Berkeley University of California Press, Berkeley, CA, p 453
Euclid: the book of balance. In: Clagett M (ed.) The science of mechanics in the middle ages. Oxford University Press, Oxford (1957)
Piaget J (1972) The child's conception of physical causality. Totowa, New Jersey; Galili I, Bar V (1997) Children's operational knowledge about weight. Int J Sci Educ 19(3):317–340
Reichenbach H (1928/1958) The philosophy of space and time. Dover, New York
Resnick R, Halliday D (1988) Fundamentals of physics. Wiley, New York, p 80; Hewitt PG (1992) Conceptual physics. Pearson, Upper Saddle River, NJ; Knight RD (2004) Physics for scientists and engineers. Pearson, Reading, MA, p 129
Ruggiero S, Cartelli A, Dupre F, Vincentini-Missoni M (1985) Eur J Sci Educ 7(12):181
Rutherford FJ (1990) Science for all Americans. Project 2061. Oxford University Press, New York; Balukovic J, Slisko J (2018) Teaching and learning the concept of weightlessness: an additional look at physics textbooks. Euro J Phys Educ 9(1):1309–7202
Stein H, Galili I (2015) The impact of operational definition of weight concept on students' understanding of physical situations. Int J Res Sci Math Educ 136:1487–1515; Stein H, Galili I, Schur

Y (2015) Teaching new conceptual framework of weight and gravitation in the middle school. J Res Sci Teach 52(9):1234–1268

Taylor BN, Thompson A (2008) The international system of units SI NIST special publication 330:52; Thompson A, Taylor BN (2008) Guide for the use of the international system of units SI. Gaithersburg National Institute of Standards and Technology NIST Special Publication, 811

Thornton W, Bonato F (2017) The human body and weightlessness operational effects problems and countermeasures. Springer International, Dordrecht; Crowell B (2006) Newtonian physics. Fullerton, California, pp 109–110

Young HD, Freedman RA (2012) University physics. Addison Wesley, New York; Reif F (1994) Understanding mechanics. Wiley, New York; Jewett JW, Serway RA (2008) Physics for scientists and engineers. Thomson Learning, Belmont, p 106

Experimentation, Impact of PER and Assessment

Experimentation in Physics Education: Should We Bother

Manjula D. Sharma

Abstract The experimental observation of gravitational waves in recent years has generated much excitement. In this instance, the interplay between theory and experiment is intriguing. The groundbreaking Eötvös experiment was undertaken around 1900, followed by Einstein's prediction. Experimental observations occurred a century later. The question arises, 'Can such experimental experiences be recreated in university and school laboratory programs'? Various phrases have been coined for such learning experiences, authentic, inquiry, to the practice of physics. Furthermore, the learning experiences are given different names, experiments, practicals and investigations; and in the current, context can be referred to as inquiry-based learning, project-based learning to STEM projects. In whatever form, curricula and pedagogies internationally are aspiring to instil the wonder of science through such pedagogies. But how can we tell if they are effective? And even more fundamentally, what does 'effective' mean? It is important to note that there are genuine challenges, from defining effective to, how to measure or evaluate. This paper seeks to provide insights by sharing two decades of research on experimentation in the Australian context. The strength of the work is that it straddles schools and universities, intertwining science disciplines, identifying commonalities and threads, binding the sciences together.

Keywords University physics education · Student practicals · Undergraduate experimental laboratories · Inquiry-based learning · Open-ended experiments

1 Introduction

The direct observation of gravitational waves in recent years has generated much excitement. One could say that there has been a sense of exhilaration. Given that gravitational waves were theorised by Einstein a century ago, and indirect observations had been made, what was so special about the direct observations? After all, the

M. D. Sharma (✉)
School of Physics, The University of Sydney, Sydney NSW 2006, Australia
e-mail: Manjula.sharma@sydney.edu.au

B. Jarosievitz and C. Sükösd (eds.), *Teaching-Learning Contemporary Physics*, Challenges in Physics Education,
https://doi.org/10.1007/978-3-030-78720-2_8

exercise was about confirming theory, not unlike school science and undergraduate physics experiments. Or was it? While a lot was known, a lot was experimental in the true sense of the word. The criticality of experimentation is captured by the following quotes.

> The principle of science, the definition, almost, is the following: The test of all knowledge is experiment. Experiment is the sole judge of scientific 'truth'.
>
> Feynman, 1963, Lectures in Physics, 1-1 (Feynman 2019)
>
> I have often had cause to feel that my hands are cleverer than my head. That is a crude way of characterising the dialectics of experimentation. When it is going well, it is like a quiet conversation with nature. One asks a question and gets an answer, then one asks the next question and gets the next answer. An experiment is a device to make nature speak intelligibly. After that, one only has to listen.
>
> George Wald, Nobel Lecture (12 Dec 1967) (Wald 1999)

Can such experiences be recreated in undergraduate laboratory programs? The quest has been on to capture such experiences in experimentation in both school and undergraduate physics experimentation. Various names have been coined, authentic, inquiry, to the practice of physics. In fact, looking back historically, the quest to emulate 'science as practised by scientists' and to capture learners' curiosity by unravelling the mysteries of nature is long standing, see, for example, (Schwab 1960; Driver 1983; Hart et al. 2000). Curricula and pedagogies have been, and continue to, earnestly design ways of engaging students through inquiries, practical work and hands-on investigations. Identifying effective practices and measuring effectiveness is paramount. With changing student demographics, increasing use of technologies and changing nature of education, the need for educators to research the practice of experimentation and hands-on investigations in school and undergraduate curricula and pedagogies continue to be fruitful and essential pursuits.

In this paper, I summarise two decades of work within the Australian context. Most involves partnerships with physics educators within the Australian Institute of Physics, the Australian Science Teachers Association and various science education networks. The work is situated within the Sydney University Physics Education Research (SUPER) group. In particular, I will focus on:

a. How do we design 'effective' experiments? How do inquiry, authentic and practice surface in the design?
b. What does it mean to measure 'student experiences of experimentation'? How do we measure? How do we interpret results to further improve experimentation?
c. Finally, if the experiences are tenuous and ill defined, how do we assess?

But first, I would like to discuss what 'effective' means? 'Effective' could purposefully have different meanings in different curriculum and pedagogical contexts. For example, in open-ended projects, 'effective' could prioritise students demonstrated competence in design, deployment and processes rather than outcomes. In specialised circuits experiments, priority could be given to dexterity with equipment and technical troubleshooting. In other experiments, priority might be given to data collection, analysis and interpretation. Another priority area is comprehension and application of uncertainty. In most cases, a range of skills and knowledge is involved, and some are selectively prioritised to build a jigsaw of learning opportunities for students. Different methodologies are needed, ranging from case studies, evaluation, to quasi-experimental methodologies. Each methodology would predicate instruments, tools, data collection and analysis. Needless to say, 'what we are striving to teach students is what we need to practice in researching student learning of experimentation'.

2 How Do We Design 'effective' Experiments?

Experiments are an integral component of both school science and undergraduate physics education (Sharma et al. 2021,2008; Hofstein and Mamlok-Naaman 2007). Designing 'effective' experiments' is not a trivial task. I will present one pedagogical tool which we have found useful in designing and/or revising school science as well as university undergraduate physics experiments. The approach is based on skills underpinning experimentations which appear in most curricula and science pedagogies, namely (1) questioning/predicting/hypothesising; (2) planning including planning data collection; (3) conducting and recording; (4) processing, analysing and evaluating; (5) reasoning, problem solving and connecting with science; and (6) concluding and communicating. These skills are shown in the first column of Table 1.

Table 1 Advancing science and engineering through laboratory learning (ASELL) schools inquiry slider; from Cornish et al. (2019)

Question	No question	Teacher provides question	Learner sharpens questions	Learner selects question	Learner poses question
Plan	No planning	Teacher provides procedure	Teacher discusses possible plans	Learner guided while planning	Learner determines plans
Conduct	Teacher conducts	Learner told how to conduct and record	Learner sharpens plan and conducts	Learner guided while planning	Learner determines plans
Analyse	Teacher analyses	Learner told how to analyse data	Teacher discusses possible analyses	Learner guided in analysis	Learner analyses data identifying trends
Reason	No problem solving	Teacher provides reasoning and links	Teacher discusses reasoning and conclusion	Learner guided in reasoning to formulate conclusion	Learner reasons to formulate conclusions
Conclude	No conclusion	Teacher writes conclusion	Learner writes conclusion	Learner guided on justifying findings and communicating	Learner justifies findings and communicates
Level of inquiry	Demonstrated Inquiry	Prescribed Inquiry	Structured Inquiry	Guided Inquiry	Open Inquiry

As one moves from the left to right of Table 1, there is a progression from directions being provided by the teacher or resources, to the learner being more self-directed. As students take responsibility of their own learning of those skills, they shift from 'being' directed' to 'self-directed'. The last row shows terms coined for the progression, such as guided inquiry. The pedagogical tool can be found in different formats with variations in the way in which the terms are used. Nevertheless, the conceptual basis is widely accepted. In the form shown in Table 1, it is called the ASELL Schools Inquiry Slider (Cornish et al. 2019; Gordon et al. 2015). The skills shown in Table 1 can readily be shaped, recast and situated within different curricula across all sciences internationally. Not all experiments lend themselves to be open-ended and not all experiments should be. Experiments will not map vertically to each level, not every student will be as adept at each skill. Hence, the Inquiry Slider provides a teaching tool, with some teachers also using it as an assessment tool. It is also a tool for programming school science and for providing variety of inquiries being offered to students.

I present two examples of the use of the inquiry slider which underpins the ASELL Schools project (ASELL Schools Link 2021). First, Science in Your Pocket (Gordon et al. 2019) in which students start off by using an APP on a mobile device to 'measure light' followed by generating their own question which they seek to investigate. Second, Vampire Power (Kota et al. 2019) in which students start of by selecting an appliance from those available and recording power readings on a spreadsheet prior to generating their own question which they investigate. Both of these integrate digital technologies and have a level of independence in the second half of the experiment. These are two-part experiment where the first part is to ease students into the 'space of the experiment' and to 'connect the whole class', and the second part is for the main experiment. The whole class activity on a spreadsheet enables the class to have a shared experience at various stages of the experiment. These experiments are normally completed within 60 min in a school science laboratory. We have run these across Australia in many schools, together with other experiments also based on the inquiry slider, see ASELL Schools project (ASELL Schools Link 2021) for more detail.

For two decades, the School of Physics has been running open-ended projects, often referred to as open inquiry, (Sharma et al. 2014) with large cohort size of around 1000 first-year undergraduate students. The projects integrate into a fairly standard first-year laboratory program. During the first four weeks, students working in teams carry out electricity/circuits experiments with 30 min each week to plan their project guided by tutors. They submit a project proposal and carry out the project in the next 4 weeks. A fleet of formative assessment tasks are deployed to support students culminating in a presentation and written report.

New research is purposefully redesigning the ways by which students engage with experiments by moving theory into appendices so that students explain their findings, rather than confirming the theory presented. We are incorporating strategic instances of 'stories' and 'colour' focusing on emotional engagement. Perhaps, the most exciting is the ways in which we are considering the teaching and learning of uncertainties and integrating digital technologies.

3 What Does It Mean to Measure or Evaluate 'Student Experiences of Experimentation'?

The words measure and evaluate are distinct. When measuring, an instrument or tool is specifically designed to measure attributes with a level of validation that the instrument is measuring what it sets out to measure. An evaluation is a broader and more diffuse scoping, often exploratory but guided by questions and a purpose. In some instances, we have evaluated while in others, we have measured.

Here, I start off by discussing the evaluation of the open-ended projects mentioned above through observations and surveys (Sharma et al. 2014). We were scoping if the learning objectives were identifiable by the students when they answered Likert scale items and through open-ended responses. We were also exploring for themes we had not anticipated. The learning objectives which ranged from students being able to undertake independent research to critical interpretation of their results, received predominantly positive ratings on the Likert scale items. The most prominent themes were 'intrinsic nature of projects' and 'their teams, including working with tutors'. These themes attracted substantive numbers of positive responses as well as fewer numbers of negative comments. The aspect which stood out the most was the role of the tutors, their support was welcome but could also be directive and intrusive. The inquiry slider (Cornish et al. 2019; Gordon et al. 2015) provides a pedagogical tool for the tutor to manage their interactions with their teams. The most pleasing aspect was that the phrase 'critical thinking' was not in the survey, but students spontaneously self-reported, critical thinking, whatever it means, and however, it was defined in their minds.

A different national Australian project designed a specific instrument, the ASELL Student Learning Experiences (ASLE) survey to measure students 'learning experiences' of a particular experiment immediately after students had completed that experiment (Barrie et al. 2015; Yeung et al. 2011). By 'learning experiences', we mean tangible pedagogical aspects of the experiments as designed by the academic in charge, as well as notions of 'experiences' from the literature on motivation. ASLE was initially trialled with 3153 chemistry students from Australia, US and New Zealand (Barrie et al. 2015), followed by 2691 students from a range of disciplines including physics (Yeung et al. 2019). The two-factor theory of motivation (Herzberg 1968; Bassett-Jones and Lloyd 2005) has been used as an interpretive framework. After checking for assumptions, exploratory factor analysis was used to extract two distinct factors: '*experiment-based motivators*' and '*course-level resources*', see Table 2.

So, what is the big deal? When correlating the items with 'overall learning experience', we find that all the items in one factor have a similar pattern. However, the pattern for 'experiment-based motivators' is distinctively different to 'course-level resources', see Fig. 1. Our analysis suggests that the items in the 'experiment-based motivators' align with student satisfaction with their experiences. Those in the ' course-level resources' appear to be more subtle; if not done well, these give rise to student dissatisfaction, but once at a certain level do not contribute to further

Table 2 ASLE items and the factor loadings [from Yeung et al. 2019]

Item numbers	Experiment-based motivators	Course-level resources
2. Laboratory skills	0.771	
1. Data interpretation	0.734	
6. Increased understanding of discipline	0.689	
3. Interest in experiment	0.687	
12. Responsibility for own learning	0.614	
10. Relevance of experiment to discipline	0.606	X
7. Background material		0.766
9. Laboratory notes		0.759
8. Demonstrator supervision		0.642
4. Clear assessment guidelines		0.661
5. Clear learning expectations	X	0.601
5. Clear learning expectations	X	0.601

satisfaction. The key message for teachers is to invest in items on the motivators to continue improving student experiences while those in the other factors will not influence student experiences after a certain level.

A final survey, the ASELL Laboratory Program Evaluation (ALPE) focusing on learning experiences of semester-long laboratory programs, has been developed and deployed with 9790 students, in physics and four other disciplines. The essence of most of the items is largely unchanged but have been edited to align with the laboratory program. A few items which are not relevant to the semester-long laboratory program have been removed and replaced with items on ethics and communication. Preliminary analysis suggests that the two factors are still consistent with robust reliabilities and loadings. The conceptual basis of the second factor is now framed around graduate qualities and capabilities. This is an important finding because it is not unusual to find that in curriculum mapping exercises, graduate qualities are mapped onto laboratory programs. Our finding that students self-report on the ALPE their experiences of graduate qualities in the laboratory program is reassuring.

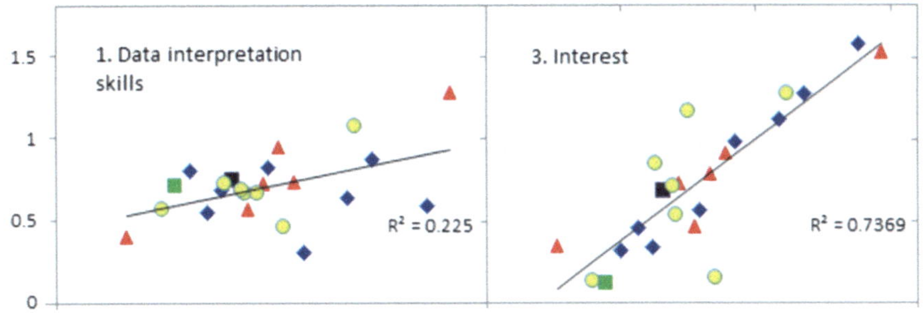

(a) Experiment-based motivators

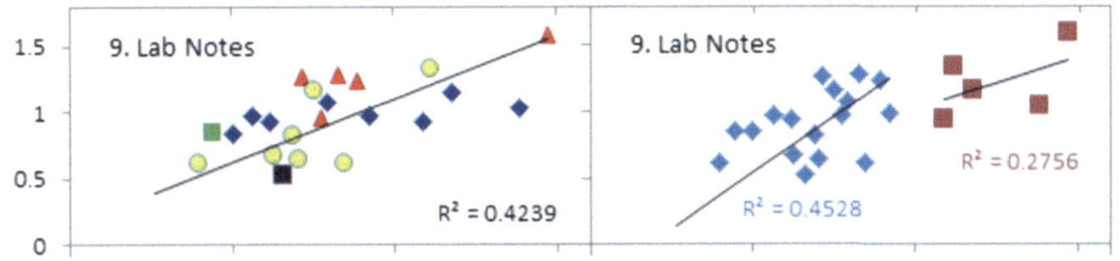

(b) Course-level resources

Fig. 1 Distinctly different patterns when correlating items with 'overall learning experience' for the two factors; **a** experiment-based motivators are different to **b** course-level resources. The data points are from different disciplines represented by different shapes, but do not influence the pattern (Yeung et al. 2019)

4 How Do We Assess Students?

This is an eternal quest. How do we assess process skills, including those which are often times referred to as soft skills? In a cross-sectional study, we gave first, second and third-year students the same task at the beginning of the year (Richardson et al. 2008). With respectable sample sizes using qualitative coding as well as marking, we found that there is a progression in the 'levels of sophistications' as students advance through their years doing experiments. In particular, their dexterity with handling equipment and technical troubleshooting improved, as did data collection and analysis skills. There was demonstrated improvement in handling of uncertainty and interpretation of their results. In other words, their experimental skills specific to physics were improving, building their jigsaw of disciplinary expertise.

Currently, we are deliberately aligning each experiment based on constructive alignment (Biggs 2003) and the ASELL Inquiry Slider (Cornish et al. 2019; Gordon et al. 2015). The skills are articulated collectively, giving rise to three assessable tasks:

Conduct and collect data.

Analyse, including uncertainties.

Interpret.

Each experiment has three learning outcomes, see Fig. 2 for an example.

TERMINAL VELOCITY

Please complete your pre-work on your eLearning account during the week of your lab session.

Learning outcomes

After completing this experiment, you should be able to:

- Undertake experiments to measure the terminal velocity of paper gliders and similar objects.
- Use logarithmic graphs.
- Explain your results, including uncertainties and their contributions.

Fig. 2 Example of the articulation of learning outcomes, aligning with inquiry slider, for an experiment

The experiments contain some well-defined, recipe-type sections while other sections are guided. There is a trajectory of development such that when the students get to the open-ended projects (Sharma et al. 2014) as discussed earlier they should be ready to undertake self-directed experimentation.

Each of the learning outcomes is assessed by an individual, but relatively low stakes, test. The first learning outcome is assessed via a practical test. There is a week when each of the 1000 students book a 40 min slot to undertake a hands-on practical test. Eight practical tests are made available beforehand. The beauty of this task is that students analyse which test will be easiest for them, and during the analysis, they invest a lot of effort and time into learning, which they would not have done otherwise. The second learning outcome is assessed via a mid-semester test which contains questions on lecture material as well as uncertainties and using spreadsheets. Students get to practice uncertainty through weekly online questions which attract a minuscule amount of marks, and all of these questions become available for students to revise prior to the mid-semester test. The third learning outcome is assessed via an individual laboratory report which is uploaded through TURNITIN and checked for plagiarism. While students work in teams of three during their sessions, for the report they need to select a section of one of their experiments for the report. They choose the experiment and particular section with an eye on the data and its analysis. Again, the selection and decision making makes them self-assess their work and see how they have developed through the semester. Each student from a team selects and re-analyses different data sets and mostly from different experiments. They also need to upload an image taken with their phone of the excerpt signed by the tutor of the raw data. The system has been running for two years now and has increased student engagement, reduced complaints and improved students ratings of the laboratories and the courses. We are yet to evaluate and/or measure other parameters. What we are also focusing on is students emotional engagement (Bhansali and Sharma 2019): an untapped avenue in most science education research.

5 Discussion and Conclusion

In conclusion, our measures have provided collective evidence that experimental programs are worthy. In particular, we as practitioners can strive to capture what enthrals experimental scientists in our undergraduate programs (Feynman 2019; Wald 1999; Schwab 1960; Driver 1983). This study shows that we can 'measure student experiences of laboratory learning' and use our measurements to iteratively improve student learning experiences (Sharma et al. 2014; Barrie et al. 2015; Richardson et al. 2008). A key challenge is engaging our colleagues in our quest as they often have good intentions but competing demands on their time. Various professional development opportunities have been designed and implemented in the Australian context (Cornish et al. 2019; Yeung et al. 2011). Our efforts are now focused on articulating with university-level key performance indicators, such as graduate attributes, so as to entrench the status and need for experimentation in the sciences. Further erosion of investment and support for experimentation must be halted. We believe that we as practitioners should embark on a campaign to gather and utilise solid evidence aligning with senior management goals to halt the erosion of genuine undergraduate experimentation.

Acknowledgements The work has been supported by many grants and undertaken by a team of students and collaborators. The author is immensely grateful to all.

References

ASELL Schools Link (2005). https://asell.org/. Last accessed 2021/01/15

Barrie SC, Bucat RB, Buntine MA, Burke da Silva K, Crisp GT, George AV, Jamie IM, Kable SH, Lim KF, Pyke SM, Read JR, Sharma MD, Yeung A (2015) Development, evaluation and use of a student experience survey in undergraduate science laboratories: the advancing science by enhancing learning in the laboratory student laboratory learning experience survey. Int J Sci Ed 37:1795–1814

Bassett-Jones N, Lloyd GC (2005) Does Herzberg's motivation theory have staying power? J Manage Dev 24:929–943

Bhansali A, Sharma MD (2019) The achievement emotions questionnaire: validation and implementation for undergraduate physics practicals. Int J Innov Sci Math Educ 27(9):34–46

Biggs JB (2003) Teaching for quality learning at university, 2nd edn. Open University Press/Society for Research into Higher Education, Buckingham

Cornish S, Yeung A, Kable SH, Orgill M, Sharma MD (2019) Using teacher voices to develop the ASELL Schools professional development workshops. Teach Sci 65:4–12

Driver R (1983) Pupil as a scientist. Open University Press, UK

Feynman R (1963) Lectures in physics 1-1. https://www.feynmanlectures.caltech.edu/. Last accessed 2019/06/30

Gordon T, Sharma MD, Georgiou H (2015) Shifting towards inquiry-oriented learning in a high school outreach program. Int J Innov Sci Math Ed 23:63–74

Gordon T, Georgiou H, Sharma MD (2019) Science in your pocket teaching science: leaving high school students to their own 'devices' while designing an inquiry-based investigation. Teach Sci 65(1):17–25

Hart C, Mulhall P, Berry A, Loughran J, Gunstone R (2000) What is the purpose of this experiment? Or can students learn something from doing experiments? J Res Sci Teach 37:655–675
Herzberg F (1968) One more time: how do you motivate employees. Harv Bus Rev 46:53–62
Hofstein A, Mamlok-Naaman R (2007) The laboratory in science education: the state of the art. Chem Educ Res Pract 8:105–107
Kota S, Cornish S, Sharma MD (2019) Switched on! Student and Teacher engagement in an electricity practical. Phys Educ 54(1):015007
Richardson A, Sharma MD, Khachan J (2008) What are students learning in practicals? A cross sectional study in university physics laboratories. CAL-Labor Int 16:20–27
Schwab JJ (1960) Inquiry, the science teacher, and the educator. School Rev 68:176–195
Sharma MD, Pollard J, Mendez A, Mills D, O'Byrne J, Scott D, Hagon S, Gribble J, Kirkup L, Livett M, Low D, Merchant A, Rayner A, Swan G, Zadnick M, Zealey W (2008) What does a physics undergraduate education give you? A perspective from Australian physics. Eur J Phys 29:59–72
Sharma MD, Mendez A, Sefton IM, Khachan J (2014) Student evaluation of research projects in a first-year physics laboratory. Euro J Phys 35:025004
Sharma MD, Mills D, Mendez A, Pollard J (2005) Learning outcomes and curriculum development in physics. http://www.physics.usyd.edu.au/super/AUTC. Last accessed 2021/01/15
Wald G (1999) Nobel lectures: physiology or medicine 1963–1970. https://doi.org/10.1142/3737
Yeung A, Pyke SM, Sharma MD, Barrie SC, Buntine MA, Burke Da Silva K, Kable SH, Lim KF (2011) The advancing science by enhancing learning in the laboratory (ASELL) project: the first Australian multidisciplinary workshop. Int J Innov Sci Math Educ 19(2):51–72
Yeung A, Cornish S, Kable S, Sharma MD (2019) What can instructors focus on when improving undergraduate science experiments? Supporting a cross-disciplinary approach Int J Innov Sci Math Educ 27:25–40

Context and Transfer: How Physics Education Research Informs Teaching and Learning

Dean Zollman

Abstract An important aspect of physics education research is understanding how students use their prior knowledge in making sense of the concepts of physics that they are studying. In these situations, the students must transfer knowledge which was acquired in a different context to a situation in physics. Much research has shown us that students rely heavily on their experiences which occurred before they studied physics when interpreting or applying the principles, while they are studying physics. Thus, they transfer to physics experiences from other formal learning, from everyday life, and from other learning environments. Sometimes this transfer is appropriate and helpful; other times it leads students down a wrong path. However, transfer *within* physics learning is equally important. Students need to be able to use knowledge from one part to learn successfully other concepts. As with transfer to physics, this type of transfer can sometimes be helpful and sometime hinder learning. This paper reviews some aspects of such transfer and includes an example in which the author needed to transfer some previously acquired knowledge to a new situation.

Keywords Transfer · Context · LEDs

1 Introduction

For over 30 years, research on teaching and learning of physics has provided a large amount of data and results about effective (and not-so-effective) learning and teaching methods. These data can provide guidance for our teaching. At the same time, teaching and learning are very complex activities, involving many variables. Thus, one size does not fit all. Nevertheless, we can gain insights to appropriate approaches to our teaching from the research.

In this paper, I will provide a view of some research that has guided me during my teaching career. In particular, I will focus on transfer of learning and how transfer is related to context. In addition to discussing how transfer applies to physics teaching

D. Zollman (✉)
Kansas State University, Manhattan, KS 66502, USA
e-mail: dzollman@phys.ksu.edu

B. Jarosievitz and C. Sükösd (eds.), *Teaching-Learning Contemporary Physics*, Challenges in Physics Education,
https://doi.org/10.1007/978-3-030-78720-2_9

and learning in general, I will provide a specific example. That example concerns something new that I learned as I was preparing the talk on which this paper is based. So, I will use my own learning as an example of the transfer process.

2 An Experiment

I will start with an experiment using the device shown in Figs. 1 and 2. It is a very simple machine. It has a column of 12 LEDs. The top LED is white. Then as you go down the column, the light emitting diodes go from red through violet. Figure 1, taken by the PER group in Udine, shows the device with all LEDs turned on.

In the USA, this apparatus costs about $170. As you turn a switch at the top, you go through all of the LEDs with color one at a time. (See the video at https://youtu.be/Qt2KIFnrdkw) In addition, the switch has one mode in which all LEDS, including the white one, are on. That is the important mode for us. The device also has labels next to each LED which states its color and the wavelength. These labels will become important.

The people who sell it seem to think that its primary use should be to show the various colors of the LEDs as shown in the video referenced above. That is a somewhat boring experiment.

When I saw this device, I immediately thought of another experiment. I could connect a variable power supply, set the switch so that all LEDs are energized and

Fig. 1 LED device with all LEDs turned on (PER Group, University of Udine.)

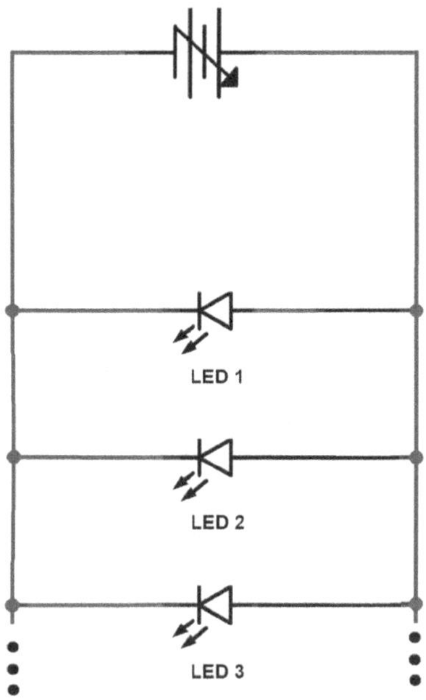

Fig. 2 Circuit diagram showing the wiring arrangement for the experiment

then vary the voltage, starting at zero volts. Watching what happens as the voltage changes would be very instructive. As we shall see, it was even more instructive than I thought at the time. That is the experiment we will do. As shown in Fig. 2, we will apply a voltage to all the LEDs simultaneously, measure the voltage as we change it, and observe the light emission form the LEDs.

My former post-doc Lei Bao, now a professor at Ohio State, had built something similar to this device many years ago. Bao's device displays the voltage digitally. Fortunately, he also built a voltage output, so I used that in this experiment. You can see what happens at https://youtu.be/uy-xTadGobY.

Typically when students describe what they saw as the voltage went from zero up to a maximum of about four volts, they note that the LEDs turned on in sequence starting with the red one and proceeding up through the violet. However, a few will also notice two anomalies. The white LED does not turn on until the blue one comes on, and one of the green LEDs does not turn on in "proper" sequence. You can view the experiment again with some hints about where to look for these anomalies in the video at https://youtu.be/r7XFx1rY8fA.

To emphasize that the bright green LED is an anomaly, Fig. 3 shows some data. As the graph demonstrates, the threshold voltage of all LEDs with a single color lies on a straight line when plotted versus wavelength. Students can obtain a good estimate of Planck's constant from that straight line. That bight green one is a problem, but I will save a discussion of it until later.

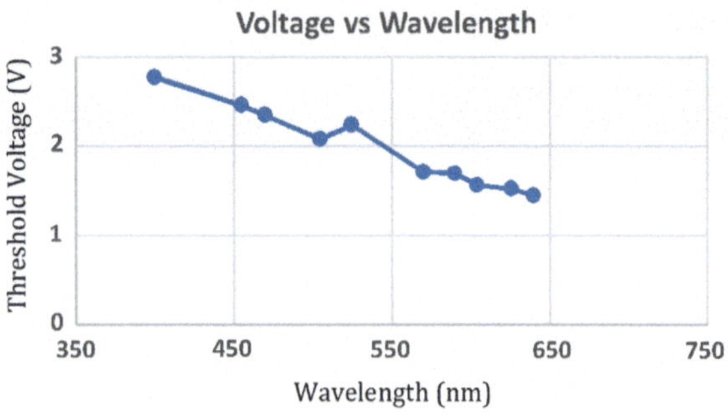

Fig. 3 Threshold voltage versus wavelength for all of the LEDs which emit light in a narrow band of the spectrum

3 The White LED

First, we will consider the strange behavior of the white LED which does not come on until the blue ones come on. A simple energy band and energy gap model for light emission from an LED is that electrons are promoted from the lower energy bands to the higher energy band by the electrical energy. When they make the transition back to the lower band, they emit light. The color of that light is related to the energy gap between the two energy bands. Using this model, we can build a model of energy bands and gaps that will emit white light. Figure 4 shows such a model which students can build using the LED spectroscopy program from our *Visual Quantum Mechanics* instructional materials (KSUPER Group, Visual Quantum Mechanics). If this model existed in a real material, the lowest energy that an electron would emit is when the electron went from the lowest energy in the higher band to the highest energy in the lower band. (Transition A in the figure.) The photon emitted would be in the red part of the spectrum. When an electron went from the highest energy in the upper band to the lowest energy in the bottom band, it would emit a blue photon. (Transition B

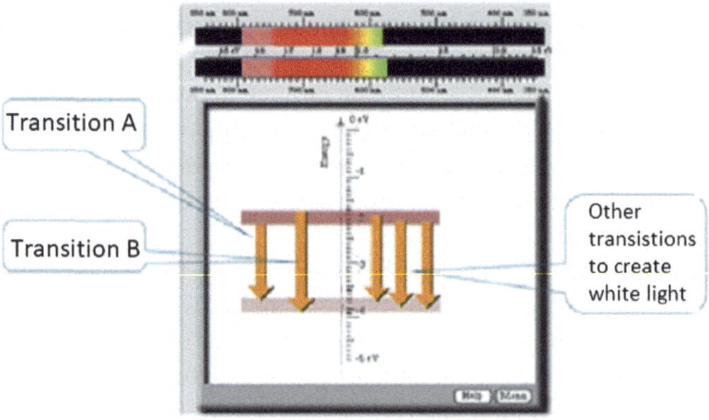

Fig. 4 Hypothetical, but incorrect, model for the emission of white light by an LED

in the figure.) Photons between these two energies would appear as electrons made other allowed transitions. Thus, this model is one that would allow for white light.

With this model, the white LED should first glow red at low voltages; then as the voltage increases, it should show the combination of red and green (yellow). Increasing the voltage even more should stimulate the emission of red, green, and blue photons, i.e., white light. However, a white LED does not behave in this way as the voltage across it increases. It turns on, emitting white light, at a specific, relatively high voltage. That is, it does not come on until those LEDs at the bottom of the device turn on.

Because this model does not work, let us review how the white LED does work. (Planinšič and Etkina 2015) The diode that is actually producing the photons is emitting blue or violet light; it is not emitting white. Sitting above this diode is a fluorescent material that absorbs some of the light and converts it to other colors—the same way fluorescent lamps work. We see a mixture of that original blue, and all the other photons of emitted by the fluorescent material. Thus, the reason the white LED does not turn on until the high voltage is because it really is a blue LED, and it needs the relatively large voltage before it can work.

4 Transfer of Learning

Before returning to the bright green LED, I will change the discussion to some research that I apply to my teaching and, as you will see, my own learning. Eventually I will show how these ideas help explain my approach to understanding that bright green LED.

The first idea goes way back to the 1960s. A group at Berkley led by Bob Karplus was developing some science teaching materials for primary schools (Karplus 1974; Karplus and Karplus 1970). Based on the work of Swiss psychologist Jean Piaget, they created an instructional model called the learning cycle. It is now a very old idea but I still use it a lot, including during my GIREP talk on which this paper is based. I will limit my discussion about it here because I have presented it many other times (Zollman 1990, 1995). A more recent idea, at least in my thinking, is transfer of learning. We know is that students build upon existing knowledge and learning. However, there are a variety of ways to look at how that building on previous learning occurs. An approach, that has served me well, was presented a seminal paper almost 20 years—"Rethinking Transfer" by Bransford and Swartz (1999). They talk about transfer in terms of what psychologists at that time thought about transfer. That is why, they were "rethinking." They were trying to get psychologists and educational researchers to think about transfer in a somewhat different way from the ideas used for most of the 20th century.

The primary change from previous ides about transfer was that knowledge is not necessarily a large coherent model that students learn in one situation then use intact when they encounter a new situation. Instead, students frequently use small pieces of knowledge from many different sources when encountering a new situation.

They assemble these pieces together in different ways for different situations and for different contexts.

One issue for education research is that it is very difficult to probe these little pieces of knowledge and how students put them together without influencing the students' thinking. When we ask students to explain something, we can cause them to put their learning together in a way that may be different from the way they would use if no one asked them about it. Thus, this way of probing students' ideas focuses on the dynamics. That is, the dynamics of the transfer of what students have learned previously and how they apply it to a situation that they have not seen before. This approach to research is quite different from taking a "snapshot" of what students know at a given time. Instead of their state of knowledge, we are probing how they build their knowledge.

I first became interested in these ideas, when one of my students, Zdeslav Hrepic, who is now a professor at Columbus State University is the US State of Georgia, was interviewing students about their conceptions on the propagation of sound. (Hrepic et al. 2005, 2010) The research methodology involved a somewhat standard misconceptions type of interview. One of the students whom Zdeslav was interviewing said, "You know, I don't know anything about this. I'm just making it up as I go." So, we started thinking about that student's statement. He did not want to say, "I don't know the answer to your questions." Instead, he thought, "You're asking the question, I must be able to say something." So, he was taking different pieces of knowledge and putting them together in a way that made sense to him and seemed to address the questions that Zdeslav was asking.

Thus, we started looking at these interviews with the idea of investigating the dynamics of transfer of learning rather than just what the student knows. We started trying to address questions such as, "What are the students constructing as they are talking during the interview?" Sanjay Rebello, now a professor at Purdue University, was a significant part of our research on transfer and also on developing instruction to help students pick the right pieces of learning to transfer. (Rebello and Zollman 2004; Rebello et al. 2005).

When thinking about what transfer means in learning physics, we expect that students can transfer their learning when they can apply concepts, solve problems, and develop new ideas by using ideas that they have learned before. However, they take the different pieces of their previous learning and put them together in new ways to address issues that they have not seen before. In this view, the different pieces of previous learning are defined very broadly. Some come from formal learning such as solving problems in the classes. But, some of the learning comes from everyday life and some from informal learning, such as going to science centers.

An example of an attempted transfer of learning is shown at https://www.gocomics.com/foxtrot/2009/01/25. The young girl, Paige, is working on an algebra problem that she cannot do. She has two simultaneous equations and must solve for x and y. She cannot do that. Her brother, Jason, reframes the problem in terms of the cost of clothing rather than in terms of x and y. Paige immediately knows the answer, but she still cannot solve the algebra problem in terms of x and y. Jason thinks she is rather weird. After all, he asked her the same problem but in a different context (clothing instead of mathematics). She solved that with ease, but she does not see the connection to her homework. She does not see how to transfer from one context to another. The important issue here is the change in context. When the change in the context was made to something with which she is familiar, she quickly sees the solution to problem. However, she cannot transfer back to a different and more abstract situation.

This is a cartoon, but it is very similar to many of the situations that we find when we are trying to help students learn. Much of the transfer research has shown similar situations to that in the cartoon. Applying our knowledge or learning to new situations is difficult.

Learning generally involves transfers in pieces. To conduct research on transfer of learning, we need to look at the dynamic situation as students are attempting the process of solving a new problem. So, the research is done in real-time situations. We must consider how things change as the student addresses a new situation rather than just a snapshot of his/her current knowledge.

Many times, students will transfer ideas that are quite inappropriate to the new situation. Then, they come to a wrong conclusion. Sometimes, these conclusions are called misconceptions. At times, however, they could be something different. An example is presented in these proceedings by Marisa Michelini. Her observations indicate that students sometimes conclude that the energy of the lines in the spectrum of gases has a one-to-one correspondence to energy of the energy levels of atoms that are emitting the light. So, if a student sees that one of the lines in the hydrogen spectrum has an energy of 2.9 eV, he or she concludes the hydrogen atom has an energy level at 2.9 eV.

We have also seen this in our research and have built one of the interactive programs in *Visual Quantum Mechanics* to address it. Our approach was that our task was to recognize that the students are partially right. A line in the spectrum indicates that the photons are being emitting at one energy. However, we needed to move the students to thinking about conservation of energy and that a transition between energy levels was needed to emit light (Escalada et al. 2005; Zollman et al. 2002).

In creating these lessons, we applied Bransford and Swartz's idea that transfer in smaller steps does seem to work well. We encourage students to bring together different small pieces of knowledge from different sources. Sometimes they need a little help so there may be some scaffolding involved. In general, taking big jumps when trying to transfer does not work, transfer in small chunks frequently does.

One way that is used in the psychology literature to characterize transfer is horizontal and vertical transfer. Horizontal transfer is applying something that you have learned but is related to situations that are slightly different from the current one. Vertical transfer is trying to modify a model or modify a way of thinking about a situation or problem. In many ways, horizontal transfer is similar to end of chapter problems, while vertical transfer is more like what we do in research.

Frequently in teaching, our goal is that we start with a novice in some subfield of physics and want them to become somewhat of an expert. Obviously, with students in any physics course, our definition of expert is different from whatever would be with a graduate student in research. Helping students apply their learning to complete some vertical, and some horizontal transfer is an effective way to bring students from the novice state to an expert state. Of course, lots of different paths could work. Research on this approach to transfer concludes that some paths work better than others. If students are asked to do too much vertical transfer too early, they can become frustrated if they are not in a position to make those changes. Likewise, if students focus entirely on a slightly different problems at the end of the chapter, they become bored. Thus, the best route to relative expertise is doing a little bit of vertical transfer, then a little bit of horizontal transfer, and repeating this process many times.

In summary, we need to balance the vertical and horizontal transfer activities. We do not expect students to do all of one and then all of the other. Almost always, we have to provide some scaffolding; students will seldom be able to do it by themselves. For example, most of my teaching materials include many short questions. The answer to one question usually leads to the next question, and so forth. This process can go on for several pages, while the students are applying little bits and pieces of their learning to new contexts. Eventually, they are able to know something quite new to them.

5 Back to the Bright Green LED

I have used LEDs in Visual Quantum Mechanics for at least 20 years (Escalada et al. 2005). The first experiment that the students perform is a slight variation on the one at the beginning of this paper. (We use an inexpensive circuit board instead of the fancy "tower.") We have written a lesson that helps students transfer what they learned about the different colors and threshold voltages and what they know about fluorescence, to be able to discover how an LED can produce white light.

However, we had never dealt with the bright green LED. I first encountered this LED when I bought the LED "tower" early in 2019. Then, I did the experiment that is described at the beginning of this paper and noticed the strange effect with the bright green LED.

To connect this observation to transfer of knowledge, I will share my experience of moving from the light emission models that I knew from previous learning to one that explained this seemingly strange behavior.

Using the interactive program that is shown in Fig. 5, we ask students to manipulate the energy bands and energy gaps. As they are doing this, they manipulate visualizations of the bands and gaps to match a hypothetical spectrum with the spectrum for each of the LEDs that they observed. I knew how that model worked, but it did not fit the observation of the bright green LED because this LED turned on at threshold voltage that was different from the one predicted by the model.

My first thought was that the wavelength that was printed on the device was wrong. The model that was correct for other LEDs was good here also. It did not seem to work only because the manufacturer had made a mistake. However, careful observation of the spectrum showed that the manufacturer's label was correct.

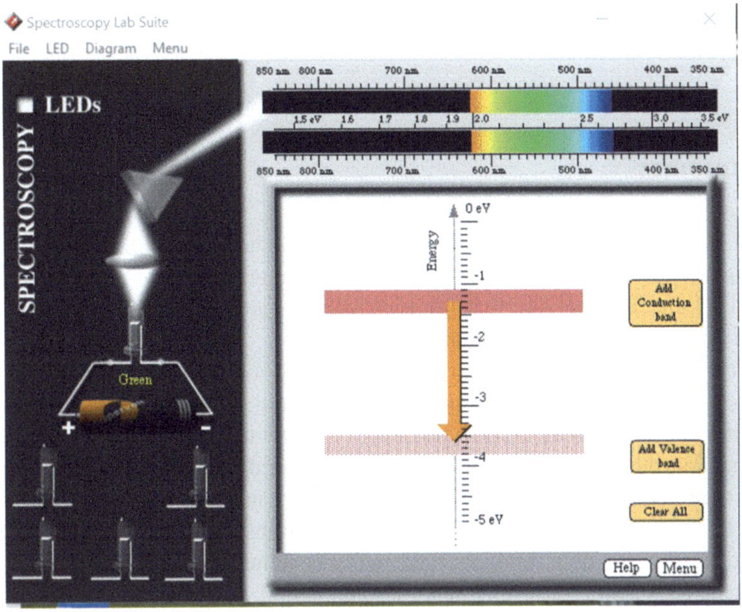

Fig. 5 LED unit in spectroscopy lab suite allows students to manipulate the energy bands. The top spectrum is the LED spectrum; the bottom one is the one created by the student energy model on the right of the screen. The program can be downloaded from (KSUPER Group, Visual Quantum Mechanics)

Only after my initial attempt to transfer my previous learning failed, I realized that I had additional information to help me understand the bright green LED. The bright green LED and the white LED turned on at the same threshold voltage. That observation started me to think about the model for emission of light from the white LED and the importance of fluorescence as part of that model.

I began to change my thinking. I transferred my knowledge about the model of the white LED to the bright green LED. This transfer was mostly horizontal. However, the two LEDs are different. The white one has a very broad spectrum covering all of visible light, while the green one has a rather narrow spectrum. Thus, I needed to change my model somewhat, that is make some vertical transfer.

The result is that my new model involves fluorescence, but the fluorescent material emits only green light. To get to this understanding, I needed to bring together pieces of knowledge involving models of "normal" LEDs which emit a light in a narrow part of the spectrum, a model of the white LEDs, and a model of fluorescent materials. Pieces from these several sources allowed me to use vertical and horizontal transfer to construct a model of emission of light for the bright green LED.

In doing some research after I had understood the bright green LED. I discovered that scientists have not yet been able to find inexpensive materials with the proper bands and gaps to emit a strong green light. The current primary method to create bright green is to use a fluorescent material to obtain green light. (Rahman 2019; Zollman and Bearden 2020).

6 Summary

When I think about transfer, I think how vertical and horizontal transfer are used by students to apply their current knowledge to new situations. As I design lessons or curricula, I check to be sure that the transfer steps are not too large and that they involve both vertical and horizontal transfer.

I can also see transfer working in my own thinking about the bright green LED. My initial attempt to use horizontal transfer failed. I could not apply my previous model for light emission from an LED to this one. In Piaget terms, I suffered a disequilibrium. Adding other pieces of knowledge, I completed a combination of horizontal and vertical transfer. My models of how LEDs work now has new components.

My takeaway message is that thinking about transfer in teaching and learning activities, and in my own thinking, have really been helpful to me. It both helps me to make better teaching–learning materials and to understand how I and my students think.

References

Bransford JD, Schwartz D (1999) Rethinking transfer: a simple proposal with multiple implications. Rev Res Educ 24:61–100. https://doi.org/10.3102/0091732X024001061

Escalada LT, Rebello NS, Zollman DA (2005) Student explorations of quantum effects in LEDs and luminescent devices. Phys Teach 42:173–179. https://doi.org/10.1119/1.1664385

Hrepic Z, Zollman DA, Rebello NS (2005) Eliciting and representing hybrid mental models. In: Annual meeting of the National Association for Research in Science Teaching

Karplus R (1974) J Science teaching and development of reasoning. J Res Sci Teach 12:165–179. https://doi.org/10.1002/tea.3660140212

Karplus E, Karplus R (1970) Intellectual development beyond elementary school. Sch Sci Math 70:398–406. https://doi.org/10.1111/J.1949-8594.1970.TB08657.X

KSUPER Group, Visual Quantum Mechanics. https://www.phys.ksu.edu/ksuper/research/vqm/

Planinšič G, Etkina E (2015) Light-emitting diodes: learning new physics. Phys Teach 53:212–218. https://doi.org/10.1119/1.4914558

Rahman F (2019) The shrinking green gap: trends in solid-state green emitters photonic spectra 53:52–59

Rebello NS, Zollman D (2004) Dynamic transfer: a perspective from physics education research. In: Mestre JP (ed) Transfer of learning: research and perspectives. National Science Foundation, Arlington, VA

Rebello NS, Zollman DA, Allbaugh AR, Engelhardt PV, Gray KE, Itza-Ortiz SF (2005) A model for dynamic transfer of learning. In: Annual meeting of the national association for research in science teaching

Hrepic Z, Zollman D, Rebello NS (2010) Identifying students' mental models of sound propagation: the role of conceptual blending in understanding conceptual change. Phys Rev ST Phys Educ Res 6:200101. https://doi.org/10.1103/PhysRevSTPER.6.020114

Zollman DA (1990) Learning cycles for a large-enrollment class. Phys Teach 28:20–25. https://doi.org/10.1119/1.2342921

Zollman DA (1995) 1996 Millikan lecture: do they just sit there? Reflections on helping students learn physics. Am J Phys 64:114–119. https://doi.org/10.1119/1.18129

Zollman D, Bearden I (2020) Determining Planck's constant with LEDs—what could possibly go wrong? Phys Educ 55:015011. https://doi.org/10.1088/1361-6552/ab525d

Zollman DA, Rebello NS, Hogg K (2002) Quantum mechanics for everyone: hands-on activities integrated with technology. Am J Phys 70:252–259. https://doi.org/10.1119/1.1435347

Teaching Thermal Phenomena and Irreversibility Through Playable Dice and Coin Toy Models

Massimiliano Malgieri, Anna De Ambrosis, Monica Di Savino, and Pasquale Onorato

Abstract In this article, we report about our progress in developing and testing a teaching–learning sequence on thermal phenomena and irreversibility for secondary school students, based on dice and coin toy models which are used both as hands-on game activity and computer simulations. Our approach can be described as laboratory-based, where the term "laboratory" is not intended as referring solely the physics laboratory, but as an integrated laboratory perspective, comprising physical experiments, game with toy models, computer simulation, group discussion, and group tasks. We report the results of a first experimentation with one class of the fourth year of secondary school (17–18 years old).

Keywords Toy model · Thermodynamics · Entropy

1 Introduction

The teaching of thermodynamics has been in past years among the most openly and widely debated areas of physics education. Poor educational results and a high incidence of alternate mental models of thermodynamics concepts have been consistently reported by many authors at the level of introductory college instruction (Loverude 2015). Based on such data, proposals of curriculum renewal, innovative teaching–learning sequences, and educational reconstructions have been advanced. In relatively recent years, the idea of a "computational" approach to the teaching of thermodynamics, heavily based on computer simulations, has gained momentum (Moore and Schroeder 1997). Many of these proposals are centered on undergraduate education, and difficulties are perceived in promoting computational thinking at the level of secondary school.

M. Malgieri (✉) · A. De Ambrosis · M. Di Savino
Department of Physics, University of Pavia, Via Bassi 6, 27100 Pavia, Italy
e-mail: massimiliano.malgieri@unipv.it

P. Onorato
Department of Physics, University of Trento, Via Sommarive 9, 38123 Povo, TN, Italy

B. Jarosievitz and C. Sükösd (eds.), *Teaching-Learning Contemporary Physics*, Challenges in Physics Education,
https://doi.org/10.1007/978-3-030-78720-2_10

Our group has been involved in research on the teaching of thermodynamics for a number of years and has been part of the drive for innovation coming from the computational perspective (Borghi et al. 2004). In recent times, we have started developing an approach to different problems in physics which is based on activities where student groups compare the results of a real low-cost experiment with those of (a) a playable dice/coin model and (b) a numerical simulation extending the stochastic model to a number of elements not easily achieved by actually performing the game with physical dice or coins. In this perspective, the design of simulations still plays a significant role, but such simulations do not take the lead from theoretical considerations but are motivated (and made easier to understand) as extensions of playable games of chance. The success of the toy model and simulation in emulating the behavior of real systems is then, only as a last step of the activity, explained in terms of a more complete microscopic theory.

This idea has been tested with problems coming from different areas of physics: radioactivity, luminescence, and thermal contact (Malgieri et al. 2018; Onorato et al. 2017), and in the present work, we used it as a central activity for a teaching–learning sequence (TLS) on thermal phenomena, irreversibility, and the second principle of thermodynamics for the secondary school. The TLS here presented and tested constituted a substantial revision and redesign of the one we discussed in Malgieri et al. (2016); our aim in particular was to increase the engagement of all students, proposing diverse activities capable of "activating" students with different skills and interests, improving the laboratory and game sessions, expanding the dimension related to the discussion of historical controversies.

2 Theoretical Background

The design of the teaching–learning sequence is based on a "three-dimensional approach" (Besson et al. 2010), involving a synergic integration of a critical analysis of the scientific content in view of its reconstruction for teaching, an overview of current teaching materials such as textbooks, and an analysis of educational research on the topic, with a particular focus on the literature on student difficulties (Moore and Schroeder 1997; Wattanakasiwich et al. 2013; Loverude 2009; Christensen et al. 2009; Leinonen et al. 2015). Within the analysis of the scientific content, we also include an analysis of foundational and historical-epistemological issues, which played a significant role in our initial design (Malgieri et al. 2016).

In the process of revision which led to the current version of the TLS, a central role was played by the construct of engagement (Fredricks et al. 2016,2004; Newmann 1992; Yonezawa et al. 2009; Cavanagh et al. 2008). Engagement has a long history in the science education literature, but recent consensus has formed that it should be regarded as a multidimensional construct consisting of three interrelated dimensions: behavioral, emotional, and cognitive (Fredricks et al. 2016). Behavioral engagement includes characteristics such as participation, effort, attention, positive conduct; emotional engagement primarily highlights positive and cooperative interactions with classmates, and self-identification with the current task or topic; cognitive engagement can be defined as the student's individual psychological investment in learning, which, if high, allows him to exert the necessary effort for comprehension of complex ideas.

According to Newmann (1992), engagement is improved in learning environments in which the tasks students are confronted with (a) are authentic; (b) give students the opportunity to take responsibility for design, execution, and evaluation; (c) provide opportunities for collaboration; (d) require diverse and forms of talents, maybe also those which are not often elicited in school; and (e) provide opportunities for fun. In Yonezawa et al. (2009) the authors highlight, with respect to the checklist proposed by Newmann, the importance allowing students to practice and discuss critical thinking with respect to the accepted norms of the discipline in a favorable, democratic environment. In Cavanagh et al. (2008), the balance between students expectations for learning before the activity and their actual capability to learn during the activity is proposed as a necessary element for student's engagement.

A main objective of the revision and redesign of our TLS was to improve as much as possible student's engagement with the sequence of activities, while preserving or improving the very good results on disciplinary learning reported for the TLS of Malgieri et al. (2016), which shared its core disciplinary structure with the one discussed in this article.

3 Description of the TLS

The TLS has a total duration of about 22 h and is structured according to the following steps: (1) motivating activity on reversibility and irreversibility; (2) microscopic time reversal and macroscopic irreversibility; (3) kinetic theory of gases and the first principle of thermodynamics, concepts of microstates and macrostates; (4) the approach to thermal equilibrium; (5) Boltzmann's definition of entropy and the second principle of thermodynamics; (6) the historical debate on the second principle of thermodynamics; (7) entropy and heat exchanges: a justification of Clausius' formula. In Fig. 1, the TLS is represented in schematic form; the steps 3, 5, and 7 have a mainly theoretical content and are mostly delivered in the form of frontal lessons; elsewhere, such teaching strategy is kept to a minimum; group tasks, pair tasks, whole class discussions, and gaming/coding activities are the prevailing instructional settings as will be discussed more in detail below.

3.1 Motivating Activity on Reversibility and Irreversibility

In the initial activity students, divided in groups, are required to produce videos for physical events which either look realistic also if the video is time reversed, and videos which do not have such property, using similar equipment in both cases. The

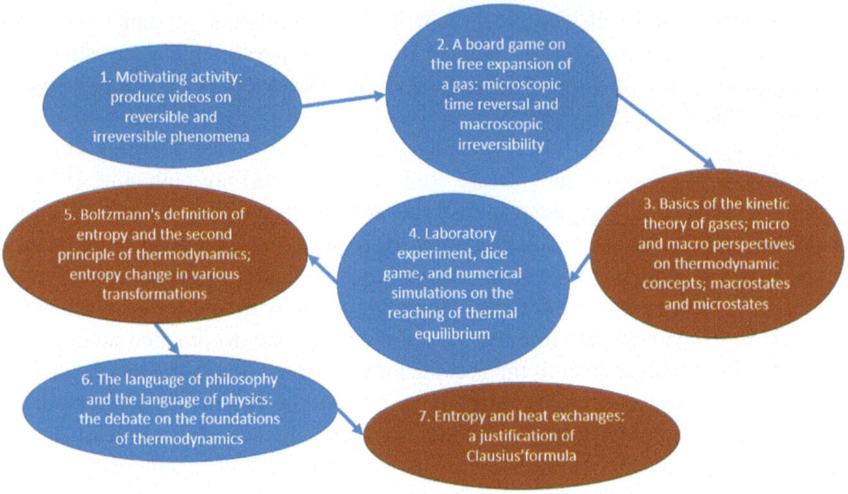

Fig. 1 Schematic representation of the TLS. Steps colored in orange have a mainly theoretical content and are delivered mostly in the form of frontal lessons; on the contrary, steps colored in cyan involve significantly or entirely group tasks, laboratories, open discussions, and other less conventional activities

activity is not preceded by an explanation of its meaning (which is discussed later) and is meant to provide students with an intuitive understanding of the concept of irreversibility.

Another aim, as discussed in Sect. 2, is to activate students' engagement on the nontrivial (but within their reach) task of producing videos which look realistic when played backwards. Students worked on several different systems (see Fig. 2), including the well-known cases of marble collisions (one-to-one and one-to-many collisions); cart collisions (elastic and anelastic); and special systems designed to provide a macroscopic analogy to the issue of energy dissipation, in which energy comes to be hidden in the microscopic world (Fig. 3).

Overall, a disciplinary goal was to provide students with the intuitive idea that irreversibility is connected to energy "spreading and sharing" (Leff 1996) either in space, or over more degrees of freedom than those it was constrained to initially. However, no connection between the spreading and sharing of energy and the concept of entropy was made at this time.

3.2 A Board Game on the Free Expansion of a Gas

In the context of a worksheet to be completed working in pairs, students play a board game meant to represent in a stylized way the free expansion of a gas. Four tokens move on a hexagonal grid cardboard according to deterministic rules (no randomness is involved) as can be seen in Fig. 4. The rules are designed in such a

Fig. 2 Screenshots from some of the videos produced by students using a recipient filled with water (Above) a ball with a similar density to the one of water floats in the recipient. The video looks credible when time reversed. (Below) a drop of ink is mixed to water. The video represents a clearly irreversible phenomenon

Fig. 3 A cart topped with oscillators (Besson et al. 2007). The oscillators can be blocked using a specially designed cover for the cart. With the oscillators blocked, the collision of the cart against a wall is (almost) elastic and the resulting video is (almost) believable when run backwards. If the oscillators are free, the collision is almost completely inelastic

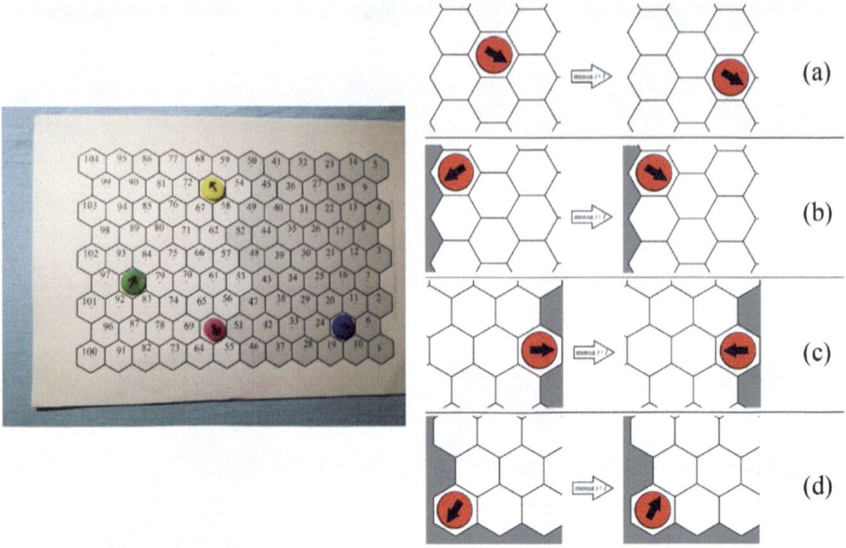

Fig. 4 Board game on the free expansion of a gas. (Left) the board and tokens (Right) the special moves representing wall collisions. Token collisions are not considered in the game

way that movement of tokens is symmetric under time reversal in the same sense as Newton's laws.

The game starts with all tokens on the left side. Students are led to reflect on the fact that, although the dynamics of the system is symmetric by time reversal, this does not mean that the two cases are equally likely that (a) the tokens, starting from half the board, expand into the other half; or that (b) being initially spread on the whole board, they spontaneously constrain themselves to half the board in a certain number of moves. In other words, the game stimulates students to reflect on the only apparent contradiction between microscopic laws that are symmetric under time reversal and the macroscopic concept of irreversibility. Proceeding on the worksheet after the game, students are led to reflect on the fact that, if the number of "tokens" approached 10^{23}, events such as the self-constraining of tokens on only one side of the box would become incredibly unlikely. Within students' worksheet, the game is also played using a random rule (with six-sided dice) for token movement. In that case, considerations of time reversal symmetry do not hold, but the general probabilistic treatment of the chance of tokens constraining themselves to half the board remains, exactly, the same. The game can be seen as an introduction to the Loschmidt paradox, which will be later discussed with students.

3.3 Laboratory Activity on Thermal Equilibrium

After a theoretical interlude in which, building on the intuitive understanding gained from the initial activities, students are introduced to a microscopic perspective on the ideal gas law, the first principle of thermodynamics, the concepts of macrostates and microstates, a new group activity is performed in the physics laboratory. There, students start with simple experiments on heat transfer, namely between water in a container and the environment, and between two water samples at different temperatures. After the experiment, students are explained a simple toy model representing thermal contact. A cardboard has two rows of squares, with different numbers of squares, which can be either filled with a coin or empty. According to the roll of two dice, coins are exchanged between the rows. The model allows to define a "toy-temperature" (Onorato et al. 2017) inspired by the kinetic definition of temperature. The toy-temperature of a row is proportional to the ratio between the number of occupied squares n_α of row α, and the total number of squares N_α of the same row:

$$\widehat{T}_\alpha \propto \frac{n_\alpha}{N_\alpha}$$

The number of squares on each row corresponds to the number of faces of two dice available in each group (e.g., a 12-sided dice and a 20-sided dice). A certain quantity of coins is initially distributed on the squares of the two rows so that one is almost filled, one almost empty. The experiment proceeds with students throwing the two dice, and moving the coin from one row to the other if the roll of the two dice corresponds to one filled and one empty square. Students repeat the process of throwing dice many times exchanging each time the positions of coins until they declare that the system is essentially "at equilibrium" (30–35 times are typically sufficient).

In the following lesson, students are taken to the computer laboratory (the activity spans 4 h in total, of which 2 in the physics laboratory and 2 in the computer laboratory) where each of the groups is given the task of writing down the abstract pseudocode for a computer program simulating the toy model for an arbitrary number of squares on the two rows and initial coins. Subsequently, the teacher helps students to write the actual code in MATLAB/Octave (no more than 20 lines of code), and perform simulations with a larger number of rows and coins observing a progressive reduction of fluctuations away from equilibrium. Finally, the teacher performs a probabilistic/statistical analysis of the toy model, and a justification of the results is given in terms of microscopic dynamics.

As discussed in detail in Onorato et al. (2017), the analysis of probabilities allows to define a "heat transfer law" governing the probability that a coin is transferred from row α to row β,

$$\begin{aligned} P(\alpha \to \beta) &= P_\alpha(C)P_\beta(E) - P_\alpha(E)P_\beta(C) \\ &= \frac{n_\alpha}{N_\alpha}\left(1 - \frac{n_\beta}{N_\beta}\right) - \frac{n_\beta}{N_\beta}\left(1 - \frac{n_\alpha}{N_\alpha}\right) \\ &= \frac{\hat{k}}{\varepsilon_0}\left(\hat{T}_\alpha - \hat{T}_\beta\right) \end{aligned}$$

where the letters C and E correspond to the state in which a square is filled with a coin or empty, respectively; the equilibrium toy-temperature can be determined,

$$\hat{T}_{\text{eq}} = \frac{N_\alpha \hat{T}_\alpha(0) + N_\beta \hat{T}_\beta(0))}{N_\alpha + N_\beta}$$

where $\hat{T}_\alpha(0)$ and $\hat{T}_\beta(0)$ are the initial toy-temperatures of the two rows; and finally, the exponential behavior of the approach to equilibrium observed also in the real experiment can be derived:

$$\hat{T}(n) = \hat{T}_{\text{eq}} + \left(\hat{T}(0) - \hat{T}_{\text{eq}}\right)e^{-n/\tau}$$

Thus, the toy model is justified as a simplification of the underlying microscopic dynamics of the real problem of heat transfer.

The toy model connects to the game of the free expansion of a gas, in the sense that allows to represent the equilibrium macrostate as the one with the highest multiplicity in microstates. The toy model also connects to the intuitive idea of spreading and sharing of energy, in that irreversibility is tied to an increase in the number of energy microstates available to the system (considering first the two objects as isolated, then in thermal contact). The activity in the computer laboratory proved to be quite engaging for students, «activating» in particular those with an interest in informatics.

3.4 Classroom Debate on the Foundations of Thermodynamics

The activity developed in three steps:

- First, in a joint lecture with the philosophy teacher, students are introduced to the language of paradoxes, starting from the paradoxes devises by ancient Greek philosophers. The goal of the lecture is to introduce students to the idea that while the scientific method signs a discontinuity between the language of physics and the one of philosophy, there is also a line of continuity, consisting in the continual use in physics of elements of language such as paradoxes and mental experiments.

- Then, students are given an individual worksheets containing original texts of physicists of the nineteenth century, debating the Loschmidt paradox, the paradox of Maxwell's demon, and the meaning of the second principle.
- Finally, the texts are openly debated in the classroom.

The discussion was very participated and, according to some of the interviewed students, contributed greatly to increase students' self-confidence. The activity was an occasion for students to practice critical thinking with respect to the accepted dogmas of the physical discipline, by taking opposite sides in historical debates; as previously discussed, this may be a key element in promoting students' engagement (Yonezawa et al. 2009). The effect of activation and increased engagement was particularly evident in the case of students with a personal interest in philosophical questions and issues.

4 Context and Data Collection

The TLS had a total duration of 26 h and was tested with a class of 23 students (12 male and 11 female) of the third year of a science-oriented high school ("Liceo Scientifico"). Lessons and activities were conducted conjunctly by the teacher of the class, who had participated in the process of revision of the TLS, and a researcher of our group.

Data for the experimentation is very rich and includes pre- and post-tests, an initial psychometric test (Raven matrices) evaluating each students' potential, individual notebooks, group laboratory reports, worksheets, the recording of all lessons, final productions of each group in the form of a PowerPoint presentation, a final online anonymous satisfaction questionnaire, and individual interviews to four students who were selected either because, according to the teacher, they had displayed highly unusual engagement in the sequence with respect to other physics lessons, or, in one case, because the student had already a pre-existing strong interest in the discipline. In this article, we discuss pre- and post-test data, and some general threads which can be detected in students' satisfaction questionnaire and interviews.

5 Data Analysis

5.1 Pre-test Data

The pre-test was composed mainly from standard items from research-validated assessments on thermal phenomena (Moore and Schroeder 1997; Wattanakasiwich et al. 2013; Loverude 2009; Christensen et al. 2009; Leinonen et al. 2015). We concentrated on students' initial understanding of basic concepts such as heat and temperature, also viewed from a microscopic perspective, and thermal equilibrium.

A hospital in a certain city keeps record of the number of child births, dividing them into male and female. Which of the following cases is more likely to happen:

a) Among the next 10 children born, 8 will be female
b) Among the next 100 children born, 80 will be female
c) Cases a) and b) are equally likely to happen.

Fig. 5 Pre-test item probing students' conceptions about the magnitude of fluctuations with respect to the mean value in samples of different size

We also included some items from standard assessments probing students' conceptions in probability and statistics (Batanero and Sanchez 2005). Student displayed several incorrect ideas well documented in the literature: for example, 18 students out of 23 defined temperature as a "measure of heat," despite having encountered the ideas of heat and temperature previously in the high school curriculum. The confusion between heat and temperatures also caused many students to refer to thermal contact as the "transfer of temperature" between bodies. Students showed difficulty also with probability and combinatorics, in particular only 6 students out of 23 correctly determined how many numbers of 3 digits can be obtained using 4 basic digits. Also, students performed very poorly with the item reported in Fig. 5, which concerns the magnitude of fluctuations in samples of different dimensions, confirming a difficulty already noticed in the literature in the context of the teaching of thermodynamics (Loverude 2009).

In this item, 17 out of 23 students answered that the two cases are equally likely, and only 3 students provided the correct answer a). As reported in Loverude, 2009 this kind of question may be very difficult even for advanced undergraduates. The very low result for this question further highlighted the importance of conducting numerical simulations of our model systems with increasing number of elements, in order to observe directly fluctuations fade away with the growing magnitude of the considered sample.

5.2 *Post-test Data*

Data from the disciplinary post-test generally provided encouraging indications, as it did in the version of 2016: although both samples were small, the disciplinary results can be considered similar. For questions which can be compared with the literature (microscopic interpretation of the compression of an ideal gas, variation of the number of accessible microstates for two items in thermal contact, variation of entropy in the same situation) results are comparable with or better than those reported in the literature for University students (Loverude 2009; Christensen et al. 2009; Leinonen et al. 2015). A majority of students seemed to have gained a reasonably good understanding of the concept of entropy, and about half the students could analyze in detail, including correct and complete conceptual explanations, the situation in which the entropy of the universe is increased due to irreversible heat exchange between an initially hot body and the environment. Difficulties persisted in the definition of temperature (7/21 students) and heat (4/21). However, to our disappointment, the misconception that the probability of large fluctuations from the mean is independent from the size of the sample seemed to be quite robust: notwithstanding our efforts, only 6/21 of students provided the correct answer in the post-test, in a question similar, although not identical to the one of the pre-test reported in Fig. 5.

A general thread in the post-test was that for items requiring an explanation, students generally were extremely prolific in writing, producing long and detailed explanations and analyses for the proposed physical situations. According to the teacher, such result was very unusual for the class. Even when those explanations were not, from a disciplinary point of view, completely correct, the willingness of students of elaborating and reporting them can be considered a sign of increased self-confidence and engagement.

5.3 The Satisfaction Questionnaire

Satisfaction questionnaires were given to students at the end of the sequence in anonymous form. They were presented as an online survey which students could complete at home. Question concerned the relation with teachers and fellow classmates, the perceived role of group work, pair work, and a general evaluation of their experience with the TLS. Raw approval rates were very high, with 60% of students providing a fully positive evaluation, and 40% stating that the experience was more on the positive side. Group and interdisciplinary activities were particularly appreciated. However, students were not necessarily, and not all, convinced that their understanding had been improved with respect to traditional teaching. Results from a key item are reported in Fig. 6.

We note that while almost all students report of having been more engaged with the TLS than during the usual lessons, and a majority reports an increased interest for the discipline, only about half of them think that the structure of the TLS was overall beneficial for their understanding. Given that we had different, independent measures of learning outcomes, which gave good results, but not significantly superior to those of experimentations in which our effort on students' engagement was not so high, we consider the result as satisfactory for our goals. Overall, students' results and questionnaire confirm the prevailing view that student's engagement is a multidimensional construct, whose reflexes on cognitive outcomes are not trivial or immediate, although it may well be a factor influencing the quality of students' learning.

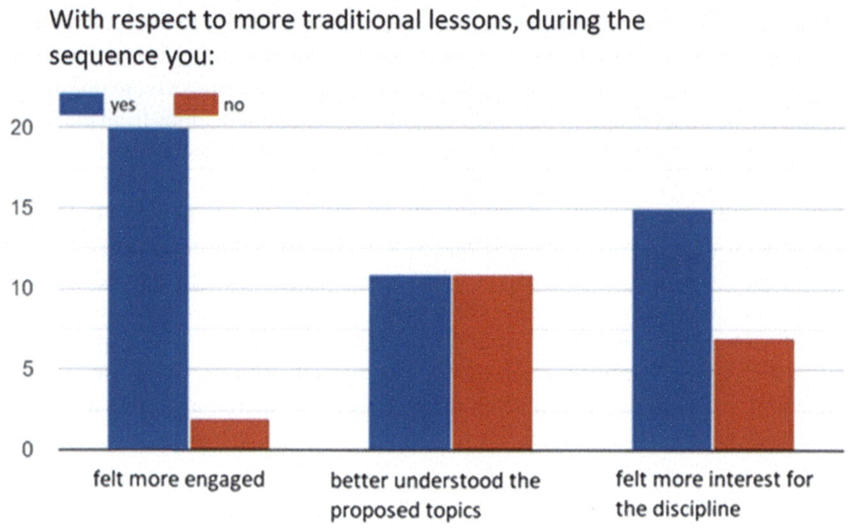

Fig. 6 Student answers to one of the items included in the satisfaction survey

5.4 Some General Results from Students' Interviews

It is possible to detect in the interviews some recurring themes: A common students' observation concerns the satisfactory experience in building the knowledge of the subject starting first from group work on games or laboratory activities. Students report that in the first stages of the sequence they could not get the general picture of what they were doing and thus concentrated on the activities with greater personal engagement and without prejudices on what the intended results should have been. However, in the end they were able to connect all what they had done in a coherent picture. One student says "[…] If one had begun with theoretical lessons, probably there would not have been so much attention, or would have felt so much involved, instead in this way, when theory was presented it connected to things we had seen already in practice, and all became simpler, so for me it was useful."

Another student observes that this approach allows to understand concepts in a deeper, and probably more long lasting way. He also remarks that working in this way, theory and disciplinary norms appear to come in response to a necessity of explanation and formalization student themselves perceive, while with the usual approach they come from authority, and are somewhat imposed to them.

Students appreciated the effort of establishing a connection between different areas of physics by adopting a microscopic perspective, and often coming back to the picture provided by mechanics. All students seem aware of the nature of physics as a connected and consistent whole, but also remark that, in the usual teaching and in textbooks, subjects and theories often appear disconnected and unrelated one with another.

One of the interviews which was particularly important for us was to a female student who had generally mediocre results in physics, but was extremely active and engaged in the TLS, and had good results in the final test. She reported the satisfaction she experienced in arriving in some case, to an explanation on her own, and highlighted the importance of the debate on historical controversies in increasing her self-confidence to tell what she thought without fearing that it could be wrong. This student reports having had previously a difficult relationship with physics, which she perceived as something which, she says, "did not correspond to me," but states that this project gave her again confidence in her possibilities, and the desire to re-approach the subject: "this project has made flourish a passion which was hidden and so… it was very useful, it brought me the desire to try again."

Some partly critical observations came from the student with a long-standing personal interest in physics. Although he also appreciated the teaching–learning sequence and most of its activities, he reported that at certain times the approach had confused him, because he was well adjusted to the usual method of having a complete and self-consistent formal presentation of the subject, which he could then understand more in depth using the textbook or other sources.

6 Conclusions

Students enjoyed the TLS. This is confirmed by worksheets, interviews, and also by the online approval questionnaire left to students after the experimentation, in which students' approval rates were very high. Part of the success of the TLS may be due to its capability to «activate», through different activities, students with different personal interests (coding, taking and reversing videos, giving a presentation, philosophical discussion…). Disciplinary results were good but, considering the smallness of both samples, they were not significantly better or worse than in the previous version of the sequence.

The effort of expanding the moments of active learning, group work, and multidimensional activity had positive effects in the self-confidence of students, their engagement in the sequence, and the general quality of their argumentation. The quality and complexity of the final productions of each student group (PowerPoint presentations) are also an element confirming such tentative conclusion.

The classroom discussion on historical controversies was also very participated and engaging for students. This was not entirely expected, but we can now confirm that 16–17 years old can understand, and become interested in, the historical debate between Boltzmann, Loschmid, Zermelo, on the meaning of the second principle, and other long-standing issues in the foundations of thermodynamics.

References

Batanero C, Sanchez E (2005) What is the nature of high school students' conceptions and misconceptions about probability? In: Jones GA (ed) Exploring probability in school. Springer, Boston, MA, pp 241–266

Besson U, Borghi L, De Ambrosis A, Mascheretti P (2007) How to teach friction: experiments and models. Am J Phys 75(12):1106–1113

Besson U, Borghi L, De Ambrosis A, Mascheretti P (2010) Three-dimensional approach and open source structure for the design and experimentation of teaching-learning sequences: the case of friction. Int J Sci Educ 32:1289

Borghi L, De Ambrosis A, Mascheretti P (2004) From university courses to teaching practice in schools: an example. In: Michelini M (ed) Quality development in teacher education and training, GIREP book of selected contributions of the second international seminar, pp 420–424

Cavanagh R, Kennish P, Sturgess K (2008) Develpment of a theoretical framework to inform measurement of secondary school student engagement with learning. In: Jeffrey P (ed) Changing climates: education for sustainable futures, 2008 annual conference of the Australian association for research in education

Christensen WM, Meltzer DE, Ogilvie CA (2009) Student ideas regarding entropy and the second law of thermodynamics in an introductory physics course. Am J Phys 77(10):907–917

Fredricks JA, Blumenfeld PC, Paris AH (2004) School engagement: potential of the concept, state of the evidence. Rev Educ Res 74(1):59–109

Fredricks JA, Filsecker M, Lawson MA (2016) Student engagement, context, and adjustment: addressing definitional, measurement, and methodological issues. Learn Instr 43

Leff HS (1996) Thermodynamic entropy: the spreading and sharing of energy. Am J Phys 64(10):1261–1271

Leinonen R, Asikainen MA, Hirvonen PE (2015) Grasping the second law of thermodynamics at university: the consistency of macroscopic and microscopic explanations. Phys Rev Special Topics-Phys Educ Res 11(2):020122

Loverude ME (2009) Student understanding of basic probability concepts in an upper-division thermal physics course. In: Sabella M, Henderson C, Singh C (eds) AIP conference proceedings, vol 1179, no 1. AIP

Loverude M (2015) Identifying student resources in reasoning about entropy and the approach to thermal equilibrium. Phys Rev Special Topics-Phys Educ Res 11(2):020118

Malgieri M, Onorato P, Valentini A, De Ambrosis A (2016) Improving the connection between the microscopic and macroscopic approaches to thermodynamics in high school. Phys Educ 51(6):065010

Malgieri M, Rosi T, Onorato P, Oss S (2018) Looking at phosphorescence with a smartphone, explaining phosphorescence with a dice toy model. Phys Educ 53(6):065016

Moore TA, Schroeder DV (1997) A different approach to introducing statistical mechanics. Am J Phys 65(1):26–36

Newmann FM (1992) Student engagement and achievement in American secondary schools. Teachers College Press, New York, NY

Onorato P, Malgieri M, Moggio L, Oss S (2017) Microscopic and probabilistic approach to thermal steady state based on a dice and coin toy model. Euro J Phys 38(4):045102

Wattanakasiwich P, Taleab P, Sharma MD, Johnston ID (2013) Construction and implementation of a conceptual survey in thermodynamics. Int J Innov Sci Math Educ 21(1)

Yonezawa S, Jones M, Joselowsky F (2009) Youth engagement in high schools: developing a multidimensional, critical approach to improving engagement for all students. J Educ Change 10(2–3):191–209

Measuring Scientific Reasoning Using the LCTSR

Maja Planinic, Ana Susac, Lana Ivanjek, Katarina Jelicic, and Karolina Matejak Cvenic

Abstract Lawson's Classroom Test of Scientific Reasoning (LCTSR) is a widely used two-tier test in physics education. The effect of three scoring methods (separate, paired, and partial credit scoring) on the person ability measures obtained with the LCTSR and on the functioning of the test itself was investigated with Rasch analysis. LCTSR was administered to a sample of 98 prospective physics teachers from Faculty of Science, University of Zagreb. Results show that statistically indistinguishable person measures were obtained with each type of scoring, but that the choice of scoring method impacted test length and targeting, and therefore also reliability and standard errors of person measures.

Keywords Scientific reasoning · LCTSR · Rasch analysis · Scoring

---

M. Planinic (✉) · K. Jelicic · K. M. Cvenic
Department of Physics, Faculty of Science, University of Zagreb, Bijenicka c. 32, 10000 Zagreb, Croatia
e-mail: maja.planinic@phy.hr

K. Jelicic
e-mail: kjelicic@phy.hr

K. M. Cvenic
e-mail: karolina@phy.hr

A. Susac
Department of Applied Physics, Faculty of Electrical Engineering and Computing, University of Zagreb, Unska 3, 10000 Zagreb, Croatia
e-mail: Ana.Susac@fer.hr

L. Ivanjek
Didactics of Physics, TU Dresden, Haeckelstraße 3, Dresden, Germany
e-mail: lana.ivanjek@tu-dresden.de

B. Jarosievitz and C. Sükösd (eds.), *Teaching-Learning Contemporary Physics*, Challenges in Physics Education,
https://doi.org/10.1007/978-3-030-78720-2_11

1 Introduction

Standard multiple-choice tests are often criticized for not providing information about the reasoning that guided students' choice of the selected answers on the test. Two-tier tests are sometimes introduced to overcome this problem. Two-tier questions consist of two parts (tiers). In the first part, there is a multiple-choice question to which students select an answer, and in the second part students select the option that best describes their reasoning in the first tier. This allows in principle not only to evaluate student answers but also the reasoning that accompanied those answers. However, that also opens a new problem of how to best score two-tier tests, since students give not only correct answers with correct reasoning, or incorrect answers with incorrect reasoning, which are both easy to classify and grade, but also correct answers with incorrect reasoning, or incorrect answers with correct reasoning, both problematic to classify and grade. Lawson's Classroom Test of Scientific Reasoning (LCTSR) (Lawson 2021) is one of the most widely used instruments for evaluation of students' scientific reasoning which is also in two-tier format. Different methods of scoring the LCTSR were compared and investigated in some studies (Xiao et al. 2018) but produced inconclusive results regarding which scoring model is optimal.

Our aim is to compare, with the help of Rasch analysis, three different methods of scoring the LCTSR (separate scoring, paired scoring, and partial credit scoring) and investigate how these different scoring methods influence students' ability estimates and the overall functioning of the test.

2 Methodology

LCTSR is a two-tier test consisting of 24 items. Items 1–20 are ten pairs of answer-reasoning items, and items 21–24 may be treated either as two pairs of hypothesis-testing items or as separate questions, since they are not in the same answer-reasoning format as the first ten pairs. In this study, items 21–24 were treated as separate items. LCTSR was administered to 98 students at Faculty of Science, University of Zagreb, who were prospective physics teachers, at the beginning of their last year of study (9th semester).

The first scoring method is separate scoring, in which each item is scored independently, and 1 point is awarded for each correct answer to either of the tiers, regardless of the student's choice of answer on the other question from the pair. It is the simplest scoring method, but often criticized because it may reward guessing on either question tier. This method preserves the original number of items, which are in this study labeled not as 1–24, as in the original LCTSR, but as 1Q (first question, labeled 1 in LCTSR), 1R (reasoning on question 1, labeled 2 in LCTSR), …, 10 Q (question labeled 19 in LCTSR), 10 R (reasoning question, labeled 20 in the LCTSR), 11, 12, 13, 14. Items labeled 11–14 are items 21–24 from the LCTSR which are not in the two-tier format.

The second scoring method is paired scoring, in which 1 point is awarded only if both questions from the pair of answer-reasoning questions are solved correctly. In this way, items 1–20 in the LCTSR are reduced to 10 items and labeled 1QR–10QR. The remaining four questions in the test (originally 21–24) are labeled 11–14 and scored separately. The paired scoring is the best way to eliminate guessing, since it is unlikely that students would guess correctly on both questions of the pair, but it may possibly also eliminate some of the partially correct answers.

The third scoring method is partial credit scoring for ten pairs of two-tier items, in which 2 points are awarded if both questions are solved correctly, 1 point is awarded if only the first question in the pair is solved correctly, and 0 points are awarded if only the second question from the pair (the reasoning question) is answered correctly, or if none of the questions is solved correctly. There could be other methods of partial scoring, some of which were investigated in other studies (Xiao et al. 2018), but we considered this one the best, since we found out in discussions with previous generations of students that they sometimes had trouble recognizing the correct reasoning option or matching their reasoning with the options that are available. This scoring method is somewhere between the very strict paired scoring and the very lenient separate scoring. However, it also reduces the total number of items in the test to 14, as in the case of paired scoring.

Data was analyzed with Winsteps software (Linacre 2021a) for Rasch analysis. Rasch analysis is an excellent tool for analyzing test functioning (Bond and Fox 2007). It produces item and person measures on the same interval scale, expressed in logit (log-odds units), provides insight into the structure and functioning of the test, and through the analysis of fit of data with the model flags problematic items and persons.

3 Results

The Rasch analysis of the LCTSR revealed that the test functioned relatively satisfactorily with each scoring method, although problems with some questions were identified. Each scoring method produced in principle a different set of measures of student scientific reasoning. However, the cross-plots of student measures suggested that the measures obtained with different scoring methods were largely the same, within the limits of their uncertainties (standard errors), as illustrated for different scoring methods in Figs. 1 and 2. From this finding, it could be concluded that it does not matter which scoring method is used (at least for this sample of students). However, the choice of the scoring method impacted some other aspects of test functioning, such as its targeting and length, as is visible from comparison of Figs. 3 and 4, which present item-person maps for the LCTSR with separate and paired scoring.

The type of scoring affected an important characteristic of the test, namely its length. The number of items was the largest with separate scoring. However, there was also inevitably a lot of dependency among two-tier items in that type of scoring, since items of the same pair generally exhibited strong mutual dependency, which is

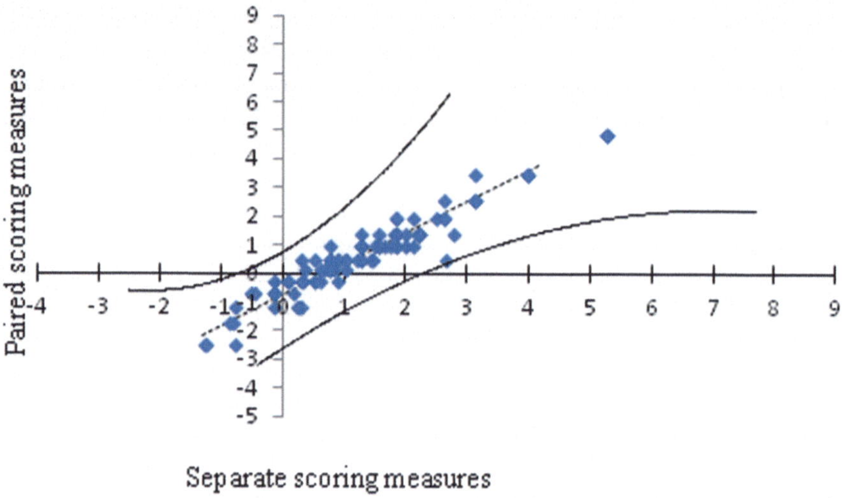

Fig. 1 Cross-plot of student measures of scientific reasoning (in logit) obtained with paired and separate scoring. The curved lines are confidence bands determined by the uncertainties of measures

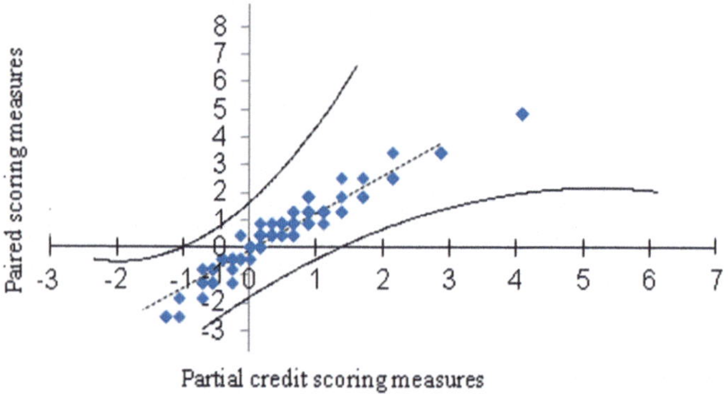

Fig. 2 Cross-plot of student measures of scientific reasoning (in logit) obtained with paired and partial credit scoring. The curved lines are confidence bands determined by the uncertainties of measures

not a favorable characteristic of a well-constructed test. The paired and partial credit scoring both reduced the number of items, which increased the standard errors (SE) of the person measures (from mean SE of 0.64 for separate scoring to 0.76 for paired scoring) and reduced reliability of person measures (from 0.73 for separate scoring to 0.68 for paired scoring).

Each type of scoring showed some advantages and some disadvantages. For example, the separate scoring, apart from being the simplest one, is also the only one that provides insight in the functioning of each tier of the question (that information

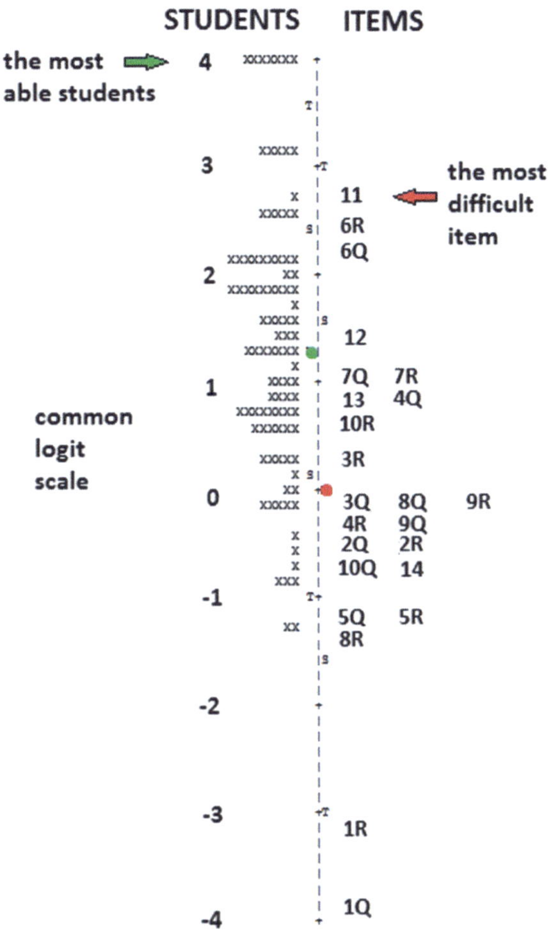

Fig. 3 Item-person map for the LCTSR with separate scoring. Each person is marked with an x. Large dots mark the position of the means of person and item distributions. Questions from the same pair are marked with the same number and letters Q (question) and R (reasoning). Items 11–14 were scored separately

is lost after pairing the items from two tiers together). This is important, so we recommend separate scoring as the first step in evaluation of a two-tier test. For example, if questions of the same pair are widely spaced in difficulty (e.g., as questions 4Q and 4R in Fig. 2) that may be an indication that students have either found a shortcut for one of the tiers or that the reasoning distracters, or the main question distracters, are poorly written. It is hard to formulate good reasoning distracters without guiding the student toward the right answer, so students sometimes have trouble choosing the correct reasoning option, even if they correctly solve the first tier of the question. The opposite is also possible, that correct reasoning option may be too obvious in some items, but that the student still does not know how to solve the first tier. In both cases, the second tier may sometimes become a task of its own, unrelated to the first tier. Also, some of the reasoning items in LCTSR showed significant misfit (e.g.,

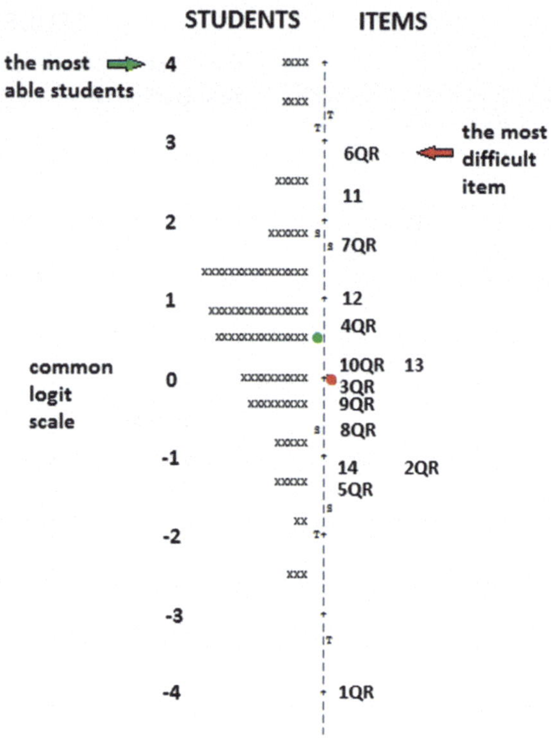

Fig. 4 Item-person map for the LCTSR with paired scoring. Each person is marked with an x. Large dots mark the position of the means of person and item distributions. Questions from the same pair are marked with the same number and letters QR. Items 11–14 were scored separately

items 4R and 5R), suggesting that students responded to these items in unexpected ways. Such problems could not be detected with paired scoring.

Large dots in Figs. 3 and 4 indicate the positions of the means of item and person distributions. Their comparison allows us to visually evaluate the targeting of the test to the sample. Pairing two-tier items together makes the test appear more difficult, and in the case of the LCTSR, better targeting of the test to this sample of students is produced.

Partial credit scoring can sometimes improve some aspects of the tests, but category functioning should always be inspected to evaluate the meaningfulness of the introduced categories. For this sample, the inspection of category functioning with the help of Category Probability Curves in Winsteps (Linacre 2021b) revealed that the middle category of 1 point in partial credit scoring was never dominant in any ability range on any item, therefore making this category obsolete—partial credit scoring functioned essentially the same as paired scoring, as the cross-plot of the obtained ability measures in Fig. 2 also indicates. This is the reason that partial credit scoring was not discussed in more detail in this analysis.

4 Conclusions

We can conclude that, at least for this sample, it does not matter which scoring method is used on the LCTSR to obtain student scientific reasoning measures. However, the scoring method can impact some other aspects of test functioning. We suggest that the simple analysis presented in this paper can help researchers to check the functioning of different scoring models on the LCTSR, or other two-tier tests, for their samples. In our analysis, all three scoring methods produced statistically indistinguishable results regarding person measures. However, different scoring methods impacted some other aspects of the test, primarily its length and targeting, which further affected reliability and standard errors of person measures. Information about the functioning of each tier can only be extracted from separate scoring—we recommend therefore separate scoring as the first step in the analysis. Paired scoring seems to give the test a more stable structure (resulting in less misfitting items) and better targeting, but since the number of items in the test is significantly reduced with this scoring method, it also reduces person reliability values and increases standard errors of person measures. We suggest that the LCTSR, which is an important instrument in physics education research, could be improved by adding more items to it and possibly combining two-tier question and reasoning pairs in one tier, which could eliminate scoring problems altogether.

Acknowledgements This work has been fully supported by the Croatian Science Foundation under the project No. IP-2018-01-9085.

References

Bond TG, Fox CM (2007) Applying the Rasch model: fundamental measurement in the human sciences. Lawrence Erlbaum Associates, Mahwah NJ

Lawson AE (2013) Lawson's classroom test of scientific reasoning. http://www.public.asu.edu/~anton1/AssessArticles/Assessments/Mathematics%20Assessments/Scientific%20Reasoning%20Test.pdf. Last accessed 2021/01/13

Linacre JM (2013) A user's guide to Winsteps Ministep Rasch-model computer programs, https://www.winsteps.com/winman/index.htm. Last accessed 2021/01/13

Linacre JM (2013) Winsteps Rasch tutorials 1–4. https://www.winsteps.com/tutorials.htm. Last accessed 2021/01/13

Xiao Y, Han J, Koenig K, Xiong J, Bao L (2018) Multilevel Rasch modeling of two-tier multiple-choice test: a case study using Lawson's classroom test of scientific reasoning. Phys Rev Phys Educ Res 14:020104

A Community of Learners on Laboratory Work. Design and Implementation of a Teacher Training Programme

Marta Carli and Ornella Pantano

Abstract This study describes the design and implementation of a professional development programme for high school teachers, aimed at improving their competences in the use of practical work for the teaching and learning of physics. The framework for designing the programme was developed based on physics education research, adopting a learning community approach. In particular, we identified five features, the effectiveness of which was tested using multiple data sources. During the programme, teachers also carried on their own action research projects. Our results suggest that all the considered features contributed to a positive change in the use of practical work. However, we argue that action research and the learning community approach were the most decisive features.

Keywords Teacher training · Community of learners · Practical work

1 Introduction

The laboratory plays a central role in physics education. When used effectively, the laboratory is the perfect setting for the development of both scientific content and abilities (Hofstein and Lunetta 2004; Etkina et al. 2006). In the last decade, the role of the laboratory in science education has been acknowledged also by several international standards and reports, both in the EU (Rocard et al. 2007; European Commission 2015; Rundgren 2018) and outside (National Research Council 2012; National Academies of Sciences 2019). However, using practical work effectively in educational contexts is not straightforward: it requires a deep awareness of both physics content and processes, together with a specific form of pedagogical content knowledge (Millar 2010; Crawford 2014). For these reasons, teachers often feel

M. Carli (✉) · O. Pantano
Department of Physics and Astronomy, University of Padua, via Marzolo 8, 35131 Padua, Italy
e-mail: marta.carli.1@unipd.it

O. Pantano
e-mail: ornella.pantano@unipd.it

B. Jarosievitz and C. Sükösd (eds.), *Teaching-Learning Contemporary Physics*, Challenges in Physics Education,
https://doi.org/10.1007/978-3-030-78720-2_12

uncomfortable in the laboratory and/or lack the competences to make practical work meaningful for learning (Nivalainen et al. 2010; Ramnarain and Hlatswayo 2018; Vanhoe and Strubbe 2014). On the other hand, few teacher training programmes address the problem specifically and provide teachers with research-based laboratory experiences (National Research Council 2006).

In the light of these considerations, we designed an in-service teacher training programme aimed at improving physics teachers' competences in the use of practical work in secondary school. In the following, after framing the research theoretically, we present the design, development and evaluation of the programme.

Our goal is to gain insights into the following research question: *What features should an in-service teacher training programme have in order to enhance the use of practical work in secondary school, and which of these features are most effective?*

2 Theoretical Background

2.1 Practical Work

Millar (2010) proposes the use of the term 'practical work' rather than 'laboratory work' to include 'any science teaching and learning activity in which the students, working individually or in small groups, observe and/or manipulate the objects they are studying', regardless of the specific setting in which they are proposed. Different kinds of activities may therefore be labelled as 'practical work'. Common categorizations have considered the purpose of a certain activity in a teaching–learning sequence (to introduce a new topic, to test a physical law, to consolidate existing knowledge), or its broader goal related to the curriculum (stimulating interest, using laboratory equipment, illustrating historical experiments). Finally, activities have also been categorized according to their 'openness' and/or their resemblance to the inquiry-based learning (IBL) model (Crawford 2014; Besson 2015; Millar et al. 2003; Woolnough and Alsop 1985; Marshall et al. 2009).

Despite the emphasis given to practical work in recent years, its role in the teaching and learning of physics, as well as the conditions that make it effective, are still a matter of debate. A point on which research is unanimous is that practical work is effective only if it is well-designed, its purpose is clear, and it is embedded in a carefully planned teaching-and-learning sequence (Millar 2010; National Research Council 2006; Besson 2015; Millar et al. 2003; Woolnough and Alsop 1985; Marshall et al. 2009; Tiberghien et al. 2000; Abrahams and Millar 2008). In fact, using practical work effectively in educational contexts is therefore not straightforward, and teachers face different kinds of difficulties in implementing practical work at school (Millar 2010; National Research Council 2006; Yalcin-Celik et al. 2017):

- Lack of preparation and support: teachers often lack personal experience in the laboratory or have weaknesses in content knowledge; on the other hand, professional development programmes rarely offer laboratory experiences that follow design principles derived from recent research.
- Absence or inadequacy of laboratory facilities and issues related to school or classroom organization.
- Poor understanding and experience of innovative instructional approaches (e.g. IBL).
- Ambiguous interpretation of national science standards: in particular, a tension is perceived between the request to cover an extensive list of topics and the amount of time required by practical work.

This picture calls for a reflection on teacher training with respect to practical work.

2.2 Teacher Training

According to Bell and Gilbert (1996), teacher training programmes should act on three dimensions: personal development (what it means to be a science teacher), social development (how to best work with others) and professional development (subject matter knowledge, pedagogical knowledge, pedagogical content knowledge). However, teacher training courses are often short and factual in nature, which does not support teachers in promoting real and long-lasting change (Gilbert et al. 2010).

A first model for successful teacher training was proposed by Adey et al. (2004), who identified the following characteristics: theoretical justification (research-based), high quality (sufficient duration, coherent methodology, intense/engaging activities, tutoring for implementation), support by the school management and sharing with colleagues. In a more recent review, DeSimone (2009) identified five 'core features' for effective teacher training:

- *Content focus*: an increased teacher's knowledge on a particular content strand has been correlated with improved student learning.
- *Opportunities for teachers to engage in active learning*, including giving and receiving feedback, observing peer teachers, engaging in discussions and reviewing student work.
- *Coherence between teacher learning and their knowledge and beliefs*: it is crucial that what is taught in professional development is coherent with national and local standards and policies, but also with teachers' beliefs.
- *Sufficient duration*: for real change to occur, time is needed, both in terms of the time span and in terms of contact hours.
- *Collective participation*: participation of teachers from the same school or grade impacts positively on the efficacy of professional development.

Concerning teachers' beliefs, a relevant class of them are *self-efficacy* beliefs, i.e. context-dependent judgements about being able to perform a particular task and obtaining the desired outcomes (Bandura 1986, 1993, 1997). Even in the context of teaching science as inquiry, the role of self-efficacy beliefs has been highlighted and their improvement has been used for the evaluation of teachers' professional development (Smolleck et al. 2006; Lotter et al. 2007,2018; Seung et al. 2019).

The importance of teacher collaboration has been emphasized recently thanks to a renewed attention to learning communities, or communities of practice (Wenger 1998). Learning communities have proven to be effective in promoting authentic innovation in physics education and in the past few years several teacher training programmes based on the learning community approach have been described (Couso 2008; Lotter et al. 2014; Singer et al. 2011). Key conditions for success of a learning community are supportive leadership, group dynamics and composition, and trust and respect (Vangrieken et al. 2017).

One further aspect that has received quite a lot of attention recently is action research. In action research, practitioners identify a problem relevant to their context, select the data they need to answer their research question, evaluate existing research, design modified classroom procedures, implement and document the activities, evaluate their project and draw conclusions in order to identify new research questions. In the context of professional development, teachers who engage in action research are supported by experts who assist them in the design step, observe them and provide feedback, discuss relevant issues and offer emotional support (Gilbert et al. 2010). Some overarching principles of action research are collaboration and reflection. In the light of this finding, a positive intercorrelation between action research and the learning community approach has been suggested (Laudonia et al. 2017; Mamlok-Naaman 2018). Recently, this interconnection has been the subject of the EU project LINPILCARE (*LINk Practitioner Inquiry* via*effective Professional Learning Communities with the results of Academic Research*), that produced precious documents and materials to be used in practitioner inquiry (LINPILCARE 2020).

3 Revised Framework for Our Teacher Training Programme

Based on a careful consideration of the literature, we identified and selected some features that constitute our renewed framework for the design of a teacher training programme focussed on the use of practical work. Our revised framework was inspired by Desimone's one but has been enriched taking into account the more recent findings described above, in particular the notion of learning community and of action research. The five features described below (and summarized in Table 1) are the ones to which our research question applies.

Table 1 Five features of our teacher training programme

Feature	Implementation
Linking content, practice and research	Focus on 'waves and their applications'. Reference to and practice on research-based activities
Action research	Each teacher wrote his/her action research plan for the year and receives feedback, support and guidance
Focus on teachers' beliefs	Pre/post-measurement of teachers' self-efficacy beliefs about teaching science as inquiry
Sufficient duration	The programme featured 13 meeting over 1 year, including 45 contact hours
Learning community	Learning community rules and style explicitly adopted. Focus on collective views beside individual views

3.1 Linking Content, Practice and Research

We decided to contextualize the reflection on the use of the laboratory into a particular content strand: we chose 'waves and their applications' (core idea PS-4 of the *Framework for K-12 science education* (National Research Council 2012)) for its centrality and transversality in the physics curriculum. All the proposed activities were based on Physics Education Research (PER). During the programme sessions, participants analysed and engaged with research-based activities, designed teaching-and-learning sequences and developed tools for putting research results into practice in the classroom.

An important source of reference for the design of our activities was the ISLE model (Etkina and Heuvelen 2007). In Investigative Science Learning Environment (ISLE), students learn physics by engaging in processes that mirror the activities of physicists when they construct and apply knowledge, and practical work is framed in terms of 'scientific abilities' that are also the focus of assessment (Etkina et al. 2006). Some ISLE ideas that we borrowed for our programme were: the identification of three different kinds of experiments (observation, testing, application); the identification of well-defined and shared learning outcomes for each activity; the use of modelling, coaching and scaffolding (guiding questions rather than 'cookbook recipes'); and the use of rubrics as a formative assessment tool and as a guide for writing laboratory reports (Etkina et al. 2002).

3.2 Action Research

We expanded and emphasized Desimone's idea of active engagement by explicitly including an action research component. At the beginning of the programme, participants wrote their own action research plan in order to apply and experiment on what they experienced during the programme. During the year, they received feedback from the colleagues and coaching from the researchers, who (1) helped the teachers formulate their research questions, (2) put each teacher in contact with relevant PER literature and (3) assisted the teachers in the design and implementation of the activities. A relevant source of inspiration for structuring action research were the results of the LINPILCARE project.

3.3 Focus on Teachers' Beliefs

As mentioned above, self-efficacy beliefs have been used recently as a measure of the effectiveness of teacher training programmes. In line with these experiences, we decided to monitor the evolution of teachers' self-efficacy beliefs on the use of practical work throughout the year using the Teaching Science as Inquiry (TSI) test (Smolleck et al. 2006), a validated instrument based on Bandura's theory of self-efficacy. The TSI considers both personal self-efficacy (SE) and outcome expectancy (OE) and explores different dimensions and levels of inquiry (National Research Council 2000).

3.4 Sufficient Duration

The programme was designed in order to guarantee sufficient time for meeting, practising, discussing, building confidence and networking. We implemented this principle by scheduling four introductory sessions in May and in September 2018, and then one session per month from October 2018 to June 2019, for a total of 13 meetings. Each meeting lasted 3.5 h, from 3:00 p.m. to 6:30 p.m.

3.5 Learning Community

The group was up as a learning community sharing common goals, rules and style. Researchers put their efforts in maintaining the community alive and active, and participants were given the opportunity to interact also outside the meetings.

4 Methods

We adopted a mixed-methods approach to evaluate our programme. Specifically, we identified four instruments aimed at providing information on different perspectives (Table 2).

Table 2 Instruments used for programme evaluation

Instrument	Information provided
Individual questionnaire	Individual views on the programme and on own practice
Focussed group interview	Collective views on the programme (five features)
Teaching science as inquiry test	Self-efficacy beliefs about using practical work at school
Individual action research reports	Effects of the programme on participants' practice

An individual questionnaire, aimed at gaining insight into individual views, was delivered at the end of the programme via the Moodle platform. It contained the following questions: *(1) How often have you used practical work this year? (2) Do you think your use of practical work has changed after the course, and how? (3) How much do you think you have improved (in physics content, use of practical work,* etc.*)? (4) What were the most useful activities during the course? (5) To what extent did you perceive the learning community approach, and you think it was useful? (6) To what extent did the programme develop the following aspects: connecting science content and practices, active involvement of participants,* etc.*? (7) How much did the course meet your expectations? (8) Comment freely on the programme.* Some of the questions were open, while others were constructed using a 4-point Likert-type scale.

In order to gather information on the community's *collective* views on the programme, we conducted a focussed group interview during the last meeting in June 2019. The discussion contained the following five questions, specifically designed to explore the five features of the revised framework: *(1) What were the advantages of discussing laboratory practices within a specific disciplinary content? How did PER results contribute to enhance the discussion on practical work? (2) In which ways engaging in action-research was useful in order to positively modify the use of practical work? (3) To what extent are self-efficacy beliefs relevant in enhancing the use of practical work? (4) To what extent was course structure relevant to promote change in your practice? (5) What was the value of setting up our group as a learning community and in what ways did this approach influence your practice?*

As mentioned above, we also wanted to monitor the evolution of teachers' self-efficacy beliefs throughout the programme. Our choice fell on the TSI test, a validated instrument that has also been used in recent literature in similar contexts. We delivered the TSI at the beginning, at the midpoint and at the end of the programme. Finally, the effects of the programme on participants' actual practice were evaluated by analysing the teachers' individual action research programmes, which they presented during the 12th session of the programme on 10 May 2019.

5 Programme Implementation

5.1 Content

The first four meetings were devoted to setting up the community, introducing the topic and writing the action research projects. During the following seven meetings, participants experimented research-based laboratories, designed their own activities, read and discussed specific didactical issues, and reflected on related aspects (e.g. assessment, use of ICT, use of out-of-school contexts). The disciplinary topics covered during the programme were mechanical waves, sound waves, ray and wave optics, and atomic spectra. In-between the meetings, collaborative online activities were proposed via the Moodle platform. These activities included: posting their own work on a forum and giving/receiving feedback, filling a database with self-produced materials and resources and a wiki for each subgroup for collaborative writing. The last two meetings were devoted to the presentation of the outcomes of individual research projects and to the final focussed group interview.

5.2 Programme Documentation

The programme was documented on the Moodle platform using a 'course journal' that contained the goals of each meeting, the description of all the activities, the links to all the relevant resources and a resumé of the reflections and issues emerged from the discussion. The Moodle page also contained the materials used during each meeting (slides, laboratory worksheets), reference to relevant research papers and other resources and reports on the results and dissemination of the project (e.g. conference slides).

5.3 Participants

The participants were 15 teachers from 11 upper secondary schools (students aged 14–19) of the Veneto region, in the north-east of Italy, where the University of Padua is located. 60% of the participants were female. Most of the participants were expert teachers (teaching experience: 5–30 years, 70% > 10 years). Their background was mostly in Mathematics (9), then Physics (3), Engineering (3) and Astronomy (1). 40% of the teachers had personal laboratory experience at university.

6 Results

6.1 *Individual Questionnaire*

70% of the teachers stated that their use of practical work has changed since they have started the course. Changes regard the use of different kinds of activities, a more open approach to practical work, and the ability to produce improved teaching materials (laboratory worksheets and assessment rubrics). The largest self-perceived improvement was in 'understanding of the role of practical work in physics education' and 'designing a laboratory activity' (average of 3.5 on a 4-point Likert-type scale). The most useful activities were 'constructing laboratory experiences' and 'engaging in research-based laboratory activities'. With respect to the programme as a whole, the most appreciated features were 'opportunities of active engagement' (3.9/4) and 'the connection between physics content and scientific practices' (3.6/4). Participants particularly appreciated the collaborative approach (3.7/4) and judged it useful (3.5/4).

6.2 *Focussed Group Interview*

The focussed group interview provided rich data about the five features. Below we analyse the five features in detail, providing a summary of the discussion.

Linking Content, Practice and Research. The chosen topic ('waves and their applications') was appreciated since it was recognized as a core idea in the physics curriculum. The physical content and didactic aspects were evaluated as equally important. The use of research-based materials was particularly relevant: one of the participants said that she 'thirsted for research-based materials in order to qualify my didactic choices' (Lucia). The reflection and practice on different kinds of experiments (observational, testing and application experiments) was particularly appreciated, since 'it provided new ways of thinking about practical work' (Sara).

Action Research. The participants recognized action research as one of the pillars of the course, even if not all of them completed their original action research plan: 'The best part of it was stopping to think about my practice in order to pose the right question' (Maria Rosa); 'Now I look at my everyday practice with a research attitude' (Alberto). Action research proved to be particularly meaningful for teachers at the beginning of their career: 'Engaging in action research gave me the opportunity to design and carry on a real research project on a real classroom' (Lucia).

Focus on Teachers' Beliefs. Working collaboratively with colleagues and sharing ideas and difficulties without the fear of judgement was decisive in fostering a positive change in self-efficacy beliefs. The active approach was a key point: 'Beliefs change if you try for yourself and you see that you like what you are doing' (Giorgio). Working on self-efficacy beliefs was particularly relevant for teachers with a background other than physics: 'I used to think, I am a mathematician, and I cannot do practical work; probably I am the wrong one. During this year I have worked on myself and now I believe that I can be comfortable in the lab and that I can promote change in my school' (Lucia). As shown by this quote, for some teachers improving their beliefs meant not only promoting personal change, but also making them think of themselves as change agents in their schools.

Sufficient Duration. The yearly duration of the programme was judged necessary in order to 'let things settle' (Maria Rosa). Participants sought even more opportunities for meeting and working together and acknowledged that 'even one year is not enough' (Alberto). They agreed that, in order to promote real change, this first year should be the first step of a long-term process.

Learning Community. The relationship environment was the key for staying and engaging in the programme. Participants particularly appreciated the possibility of meeting and interacting with colleagues from different schools and backgrounds, which gave them the possibility to experience new dynamics and see themselves in different roles: 'It was good to experience a wider network of relationships beyond the one in our schools' (Alberto). On the other hand, the presence of two teachers per school, which was encouraged, was appreciated and recognized as 'a seed to start a learning community in each school' (Francesco). One criticality was the difficulty of collaborating online: some collaborative tasks (e.g. wikis) were perceived as too demanding by people who were not used to this way of collaborating. Participants also suggested possible upgrades, such as introducing peer observation and microteaching activities. Finally, the learning community was seen as 'a powerful strategy for reinforcing the relationship between schools and university' (Giorgio).

6.3 The 'Teaching Science as Inquiry' Test

The pre-course administration of the TSI showed that the participants already held a fair self-efficacy belief at the beginning of the course (SE: 3.58/5.00, OE: 3.37/5.00). The overall improvement between the pre- and post-course administration was small (SE: + 0.14, OE: + 0.08), but individual differences were highly variable. Interestingly, the ones who improved more in the TSI were also the ones who conducted a more complete action research plan. The dimensions of inquiry where the largest overall improvement was observed were 'learner gives priority to evidence in responding to questions' (+ 0.28) and 'learner formulates explanations from evidence' (+ 0.20). Concerning the levels of inquiry, an overall increase towards a higher learner self-direction was observed (+ 0.28).

6.4 Individual Action Research Reports

All the teachers presented some results according to their initial research plans, though with variable quality and depth of the research. The majority presented single actions or improvements, but a subgroup presented complete and documented research reports. We briefly describe two examples.

Lucia, one of the youngest teachers in the group, tested the use of observational experiments in the context of mechanical waves, comparing a research-based laboratory with a traditional laboratory performed in a parallel classroom, both taught by herself. She evaluated her research using three different rubrics (her observation, an external observer's observation, and the students' self-evaluation), the analysis of students' laboratory reports and a test administered three months after the laboratory. Her results support the effectiveness of the research-based laboratory. After the experimentation, she refined her research question as 'how can I re-design the lab in order to involve all the students, including the weakest ones?'. Lucia's improvement in the TSI was + 0.93 for SE and + 0.58 for OE.

Maria Rosa is an experienced teacher (teaching experience > 20 years) whose research question was 'how can I design an IBL-based teaching–learning sequence on atomic spectra in a limited span of time, and how can I evaluate its effectiveness?'. Her proposal included a pre-test, a 2-h research-based laboratory, a 2-h post-laboratory tutorial and a post-test. She assessed both students' performance and their degree of participation using specifically designed rubrics. The classroom performance in the post-test was highly satisfactory, and the students who actively took part in laboratory group work were the ones who obtained the best results. Maria Rosa's largest improvement in the TSI was in OE (+ 0.63).

7 Discussion and Conclusions

Based on these results, we can conclude that all the features considered in our programme contributed to a positive change in the use of practical work for in-service physics teachers.

We argue that action research and the learning community approach were the most decisive features. Teachers appreciated to be actively engaged, which confirms Desimone's core feature; however, engaging in action research was particularly effective since it impacted the teachers' *view* of their teaching, fostering a more evidence-based attitude, and boosted the development of positive self-efficacy beliefs. On the other hand, participants highlighted the importance of a non-threatening, inspiring, trustful and collaborative context in which to grow as teachers and persons, which confirms learning communities as an ideal setting for teachers' professional development. Even the long timespan and high number of contact hours emerged as necessary, confirming previous findings. It became even clearer that real change is a long process, that cannot be achieved only with short, factual interventions but requires time to work 'slowly but authentically' on oneself. The importance of linking content, practice and research was also reaffirmed. Concerning the fifth feature (focus on self-efficacy beliefs), our results are not fully aligned between the TSI and the interview. We argue that more time and more opportunities for classroom practice, possibly in the form of action research, are needed to impact teachers' beliefs significantly, which is supported by the fact that the teachers whose beliefs improved more significantly were the ones who were more actively engaged in action research. This will be the subject of further research and a priority for the continuation of the programme.

In fact, in the light of this experience, we have continued the programme for another year, in which the efforts of the community have been focussed on the design of teaching–learning sequences and their experimentation in the classroom. In order to facilitate the teachers in the process, we have reinforced activities such as co-planning, microteaching, peer observation and giving/receiving feedback. The results of the two-year experimentation will be described in paper which is currently in preparation. We also plan to optimize and further validate our framework by re-proposing the programme with a new group. This action will also have the effect of enlarging the basis of schools reached by the programme. Teachers who took part in the first run could act as coaches for the new colleagues. We expect that, in the long-term, our research will contribute to produce a research-based, experimentally validated model for in-service teacher training in physics education, promote a scientific (experimental) approach to teaching, and validate a model for effective collaboration between the university and schools, where 'COLLABORAtive' teachers act as change agents in their schools, and their schools become hubs for effectively innovating physics education in the territory.

References

Abrahams I, Millar R (2008) Does practical work really work? A study of the effectiveness of practical work as a teaching and learning method in school science. Int J Sci Educ 30(14):1945–1969. https://doi.org/10.1080/09500690701749305

Adey P, Hewitt G, Hewitt J, Landau N (2004) The professional development of teachers: practice and theory. Kluwer Academic, Dordrecht

Bandura A (1986) Social foundations of thought and action: a social cognitive theory. Prentice-Hall, Englewood Cliffs

Bandura A (1993) Perceived self-efficacy in cognitive development and functioning. Educ Psychol 28(2):117–148. https://doi.org/10.1207/s15326985ep2802_3

Bandura A (1997) Self-efficacy: the exercise of control. W. H. Freeman and Company, New York

Bell B, Gilbert JK (1996) Teacher development: a model from science education. Falmer, London

Besson U (2015) Didattica della fisica. Carrocci, Rovigo

Couso D (2008) Authentic collaboration: a promising paradigm for physics education reform. Int Newslett Phys Educ (ICPE, IUPAP) 56:7–9

Crawford BA (2014) From inquiry to scientific practices in the science classroom. In: Lederman NG, Abell SK (eds) Handbook of research on science education. Routledge, Abingdon, pp 515–541. https://doi.org/10.4324/9780203097267.ch26

DeSimone LM (2009) Improving impact studies of teachers' professional development: toward better conceptualizations and measures. Educ Res 38(3):181–199. https://doi.org/10.3102/0013189X08331140
Etkina E, Van Heuvelen A (2007) Investigative science learning environment—a science process approach to learning. In: Redish EF, Cooney P (eds.) Physics research based reform of university physics. American Association of Physics Teachers, College Park, pp 2–48
Etkina E, Van Heuvelen A, Brookes DT, Mills D (2002) Role of experiments in physics instruction—a process approach. Phys Teacher 40(6):351–355. https://doi.org/10.1119/1.1511592
Etkina E, Van Heuvelen A, White-Brahmia S, Brookes DT, Gentile M, Murthy S, Rosengrant D, Warren A (2006) Scientific abilities and their assessment. Phys Rev Special Topics Phys Educ Res 2:020103. https://doi.org/10.1103/PhysRevSTPER.2.020103
European Commission (2015) Science education for responsible citizenship. Report to the European commission of the expert group on science education. Publications Office of the European Union, Luxembourg
Gilbert JK (2010) Supporting the development of effective science teachers. In: Osborne J, Dillon J (eds) Good practice in science teaching: what research has to say. Open University Press, Maidenhead, pp 274–300
Hofstein A, Lunetta VN (2004) The laboratory in science education: foundations for the twenty-first century. Sci Educ 88:28–54. https://doi.org/10.1002/sce.10106
Laudonia I, Mamlok-Naaman R, Abels S, Eilks I (2017) Action research in science education—an analytical review of the literature. Educ Action Res 26(3):480–495. https://doi.org/10.1080/09650792.2017.1358198
LINPILCARE (LINk practitioner inquiry via effective professional learning communities with the results of academic research) homepage. http://www.linpilcare.eu. Last accessed 2020/12/29
Lotter C, Harwood H, Bonner J (2007) The influence of core teaching conceptions on teachers' use of inquiry teaching practices. J Res Sci Teach 44(9):1318–1347. https://doi.org/10.1002/tea.20191
Lotter C, Yow JA, Peters TT (2014) Building a community of practice around inquiry instruction through a professional development program. Int J Sci Math Educ 12:1Y23. https://doi.org/10.1007/s10763-012-9391-7
Lotter C, Thompson S, Dickenson TD, Smiley WF, Blue G, Rea M (2018) The impact of a practice-teaching professional development model on teachers' inquiry instruction and inquiry efficacy beliefs. Int J Sci Math Educ 16(2):255–273. https://doi.org/10.1007/s10763-016-9779-x
Mamlok-Naaman R (2018) Using the action research rationale to enhance the creation of teachers' professional learning communities (PLCs). Action Res Innov Sci Educ 1(1):27–32. https://doi.org/10.12973/arise/94201
Marshall JC, Smart J, Horton RM (2009) The design and validation of EQUIP: an instrument to assess inquiry-based instruction. Int J Sci Math Educ 8:299–321. https://doi.org/10.1007/s10763-009-9174-y
Millar R (2010) Practical work. In: Osborne J, Dillon J (eds) Good practice in science teaching: what research has to say, 2nd edn. Open University Press, Maidenhead, pp 108–134
Millar R, Tiberghien A, Le Maréchal J-F (2003) Varieties of labwork: a way of profiling labwork tasks. In: Psillos J, Niedderer H (eds) Teaching and learning in the science laboratory. Kluwer Academic, Dordrecht, pp 9–20. doi: https://doi.org/10.1007/0-306-48196-0_3
National Academies of Sciences (2019) Engineering, and medicine: science and engineering for grades 6–12: investigation and design at the center. The National Academies Press, Washington
National Research Council (2000) Inquiry and the national science education standards: a guide for teaching and learning. The National Academies Press, Washington
National Research Council (2006) America's lab report: investigations in high school science. The National Academies Press, Washington
National Research Council (2012) A framework for K-12 science education: practices, crosscutting concepts, and core ideas. The National Academies Press, Washington

Nivalainen V, Asikainen MA, Sormunen K, Hirvonen PE (2010) Preservice and inservice teachers' challenges in the planning of practical work in physics. J Sci Teacher Educ 21(4):393–409. https://doi.org/10.1007/s10972-010-9186-z

Ramnarain U, Hlatswayo M (2018) Teacher beliefs and attitudes about inquiry-based learning in a rural school district in South Africa. S Afr J Educ 38(1):1–10. https://doi.org/10.15700/saje.v38n1a1431

Rocard M, Csermely P, Jorde D, Lenzen D, Walberg-Henriksson H, Hemmo V (2007) Science education now. A renewed pedagogy for the future of Europe. Office for Official Publications of the European Communities, Luxembourg

Rundgren C-J (2018) Implementation of inquiry-based science education in different countries: some reflections. Cultural Stud Sci Educ 13:607–615. https://doi.org/10.1007/s11422-016-9787-8

Seung E, Park S, Lee M-A (2019) The impact of a summer camp-based science methods course on preservice teachers' self-efficacy in teaching science as inquiry. J Sci Teacher Educ 30(8):1–18. https://doi.org/10.1080/1046560X.2019.1635848

Singer J, Lotter C, Feller R, Gates H (2011) Exploring a model of situated professional development: impact on classroom practice. J Sci Teacher Educ 22:203–227. https://doi.org/10.1007/s10972-011-9229-0

Smolleck LD, Zembal-Saul C, Yoder EP (2006) The development and validation of an instrument to measure preservice teachers' self-efficacy in regard to the teaching of science as inquiry. J Sci Teacher Educ 17(2):137–163. https://doi.org/10.1007/s10972-006-9015-6

Tiberghien A (2000) Designing teaching situations in the secondary school. In: Millar R, Leach J, Osborne J (eds) Improving science education: the contribution of research. Open University Press, Buckingham, pp 27–47

Vangrieken K, Meredith C, Packer T, Kindt E (2017) Teacher communities as a context for professional development: a systematic review. Teach Teach Educ 61:47–59. https://doi.org/10.1016/j.tate.2016.10.001

Vanhoe H, Strubbe K (2014) Chemistry education in Flanders: from a strictly concept based curriculum towards a more inquiry-based curriculum. Research report, Department of Chemistry, University of Jyväskylä, Finland

Wenger E (1998) Communities of practice: learning, meaning, and identity. Cambridge University Press, New York

Woolnough BE, Alsop T (1985) Practical work in science. Cambridge University Press, Cambridge

Yalcin-Celik A, Kadayifci H, Uner S, Turan-Oluk N (2017) Challenges faced by pre-service chemistry teachers teaching in a laboratory and their solution proposals. Eur J Teach Educ 40(2):210–230. https://doi.org/10.1080/02619768.2017.1284792

A Study on Engineering Freshman Conceptual Understanding of Newtonian Mechanics

Onofrio Rosario Battaglia and Claudio Fazio

Abstract Force concept inventory is a multiple-choice questionnaire commonly used to assess students' conceptual understanding of Newtonian mechanics. We here show that a cluster analysis method can be used to study student answers to the force concept inventory to investigate their understanding of Newtonian mechanics and provide new insights into the use of the force concept inventory. We identified groups of students characterized by similar correct answers as well as by non-correct answers to the questionnaire, whose analysis allowed us to highlight student misconceptions/non-normative conceptions. Such an analysis of student answers gave us insights into the relationships between the student ideas about the force concept and their ability to correctly answer questions involving the first and second Newton's law.

Keywords Cluster analysis · Engineering freshmen · Force concept inventory · Newtonian mechanics

1 Introduction

Problem-solving has been placed at the top of the list for desired learning outcomes in engineering education programs (Brophy et al. 2008). Disciplinary conceptual understanding is recognized as a relevant competence for the development of problem-solving skills (National Research Council 2012). Engineering students' knowledge of foundational principles in many physics fields is often weak and marked by misconceptions (Steif and Dantzler 2005; Steif and Hansen 2006; Prince et al. 2012). These foundational principles are complex and involve multiple interconnected knowledge elements that must be understood both individually and with each other. It has been

O. R. Battaglia (✉) · C. Fazio
Dipartimento di Fisica e Chimica - Emilio Segrè, Università degli Studi di Palermo, Palermo, Italy
e-mail: onofriorosario.battaglia@unipa.it

C. Fazio
e-mail: claudio.fazio@unipa.it

B. Jarosievitz and C. Sükösd (eds.), *Teaching-Learning Contemporary Physics*, Challenges in Physics Education,
https://doi.org/10.1007/978-3-030-78720-2_13

pointed out that even advanced engineering students make errors when asked fundamental questions about physics key concepts. This leads to "get beyond the symptoms to the causes" of this phenomenon (Streveler et al. 2008) by deepening research that identifies students' conceptual difficulties and mainly trying to understand their origin.

Conceptual inventories (CIs) are a class of assessment instruments designed to measure student conceptual understanding and common student errors (Steif and Hansen 2007) and are commonly used in engineering and science education. They are highly focused on a small set of key concepts within a limited academic content domain, such as the concept of force in Newtonian mechanics (Hestenes et al. 1992), selected concepts in statics (Steif and Dantzler 2005), heat and energy concept (Prince et al. 2012), and fluid mechanics (Erceg et al. 2019). A recent paper (Jorion et al. 2015) pointed out that educators often use assessment instruments without sufficient evidence that a structured approach has been applied to validate inferences about student thinking. The paper provides an analytic framework to evaluate instrument validity and make more guaranteed uses and interpretations of its outcome scores.

Among the many research-based multiple-choice tests, the force concept inventory (FCI) (Hestenes et al. 1992) has been perhaps the most widely used for evaluating students' conceptual understanding in the domain of introductory mechanics as well as to analyze in detail student conceptual knowledge in the topic of Newtonian mechanics (Hestenes and Halloun 1995). A large body of research has been developed to validate the instrument, as studies of its reliability (Lasry et al. 2011), applications of item-response theory (Wang and Bao 2010; Morris et al. 2012), Rasch analysis (Planinic et al. 2010), and two studies that use factor analysis (Huffman and Heller 1995; Scott et al. 2012). Moreover, FCI has been used to investigate the effectiveness of various teaching methods (Hake 1998). Instructors and researchers are often interested in considering the correctness of the students' answers on the FCI or their normalized gains after a teaching intervention. Moreover, they are interested in using FCI as a "diagnostic instrument," i.e., to know details about relationships among student answers to the different questions and what these answers might indicate about their understanding.

FCI questions come from a systematic investigation of student difficulties in understanding a particular concept (Halloun and Hestenes 1985). These often rely on detailed interviews to identify common models that students may form before, during, and after instruction and that produce correct and incorrect responses (Hestenes and Halloun 1995). The analysis of such wrong answers (sometimes called alternative or common-sense conceptions, misconceptions or non-normative answers) provided a rich body of knowledge about how students correctly or incorrectly reason in a relevant field of physics.

Some authors (Bao et al. 2002; Bao and Redish 2006) used model analysis to analyze students' responses to multiple-choice tests by predefined mental models. Moreover, they find the probability that students use each of these models in different contexts (Bao and Redish 2006). A recent paper (Brewe et al. 2016) describes a new methodology for carrying out network analysis on responses to multiple-choice conceptual inventories. This method is used to identify modules of non-normative responses which supply information about alternative reasoning strategies that can help classroom instructors in using multiple-choice assessments as diagnostic instruments.

In this paper, we apply a cluster analysis (ClA) method (the k-means algorithm) (Everitt et al. 2011) to study students' responses to FCI questions to identify reasoning patterns based on Newtonian conceptions as well as alternative conceptions (misconceptions/non-normative conceptions). The application of ClA techniques is common in many fields, including information technology (Cowgill and Harvey 1999), biology (Ott 1999), medicine (Allen and Goldstein 2013), econophysics, and market research (Mantegna 1999). These techniques allow the researcher to locate subsets or clusters within a set of elements of any nature that tend to be homogeneous based on some criteria. The results of the analysis should reveal high homogeneity within each subset (intra-cluster) and high heterogeneity between subsets (inter-clusters).

Our study aims to show that the cluster analysis method can be used to analyze student understanding supplying useful details about their reasoning strategies providing insights into the use of FCI as a diagnostic instrument. More detail of the analysis can also be found in Fazio and Battaglia (2019) and Battaglia and Fazio (2021).

2 Method

2.1 The Sample

FCI was administered to a sample composed of 148 freshman engineering students at the beginning of a workshop proposed by the authors as an additional/optional activity introducing their first physics degree course.

Usually, freshman engineering students come from different kinds of high schools, each of them characterized by a specific curriculum for the quantity/quality of physics courses involved. As a consequence, for many years, engineering degree courses have been organizing introductory workshops focusing on some disciplinary areas.

We proposed a workshop focusing on the Newtonian concepts, and in this study, we report only the results obtained by the pre-instruction test.

It is worth noting that we excluded from our analysis all the students who did not answer more than 20% of the items. So, we analyzed a sample composed of 116 students made 73.3% male and 26.7% female.

Table 1 Questions included by the authors in the different subtests

Subtest	Questions
SubA	Q3 Q6 Q7 Q8 Q9 Q10 Q12 Q17 Q21 Q22 Q23 Q24 Q25 Q26 Q27
SubB	Q5 Q11 Q13 Q18 Q29 Q30
SubC	Q4 Q15 Q16 Q28
SubD	Q1 Q2 Q14 Q19 Q20

2.2 The Questionnaire

FCI comprises 30 multiple-choice questions, and the paper of 1995[1] includes a table in which Newtonian concepts are classified. The table reports a decomposition of the force concept into six fundamental conceptual dimensions.

In a recent paper (Ding and Caballero 2014), the authors sorted FCI questions into five concept categories (Kinematics, Newton's First Law, Newton's Second Law, Newton's Third Law, and Force Identification), using the Hestenes' original conceptual dimensions. Each question is placed in only one category. It is worth noting that the proposed categorization comes from the researchers' judgment of the concepts covered by the questions and not as the result of internal correlations or factor analysis based on student data. However, a previous paper (Scott et al. 2012), regarding an exploratory factor analysis of an FCI data set, partially confirms such a categorization. The variation in the data set is characterized as being due to the effect of one or more underlying factors and some inherent noise.

It is noteworthy that many FCI questions do not cleanly poll a single concept. For example, a question about the second law also involves the basic kinematics concepts and the identification of forces. However, the majority of experts agree that the FCI displays the conceptual coherence on which the Newtonian concept of force is built.

We categorized the 30 FCI questions in the following four categories or subtests.

- SubA including 15 questions assessing student ability to predict trajectories as well as the ability to explain these using the first and/or the second Newton's law.
- SubB including questions assessing student ability to identify forces acting on bodies in different dynamical or static situations.
- SubC including questions assessing student ability to apply Newton's third law in dynamical or static situations.
- SubD including questions assessing student ability to correctly use kinematics quantities and laws.

Table 1 reports the questions belonging to each subtest.

[1] http://modeling.asu.edu/RE/Research.html.

3 The Results

3.1 Score Distribution and Item Analysis

We first report some statistical results. Distribution of scores for the whole sample is shown in Fig. 1. The average FCI score is $M = 8.8$, corresponding to 29.2% of correct answers, and the standard deviation is $\sigma = 4.0$.

Figure 1 suggests that the majority of our student sample, at the end of secondary school, can be still classified as non-Newtonian. The average FCI score is well below 60% of correct answers, suggested by FCI authors (Hestenes et al. 1992; Hestenes and Halloun 1995; Hake 1998) as a threshold for the student development of Newtonian thinking. Only three of our students obtain a score above 60%.

We first performed the analysis of the student answers to the different questions by evaluating their difficulty indexes and concentration factors (Bao and Redish 2001). Figure 2 shows the difficulty index of the FCI questions, Qj ($j = 1 \ldots 30$), using the proportion of students who answer a question correctly as a rough measure of difficulty: i.e., $P(Qj) = Nj/S$, where Nj is the number of correct responses to the questions Qj and S is the total number of students taking the questionnaire.

The concentration factor, C_j ($j = 1, \ldots 30$) evaluates how the students' responses are distributed among the different responding options presented by each item (Bao and Redish 2001). It is defined as

$$C_j = \frac{\sqrt{m}}{\left(\sqrt{m} - 1\right)} \times \left(\frac{\sqrt{\sum_1^m {n_i}^2}}{N} - \frac{1}{\sqrt{m}}\right) \quad (1)$$

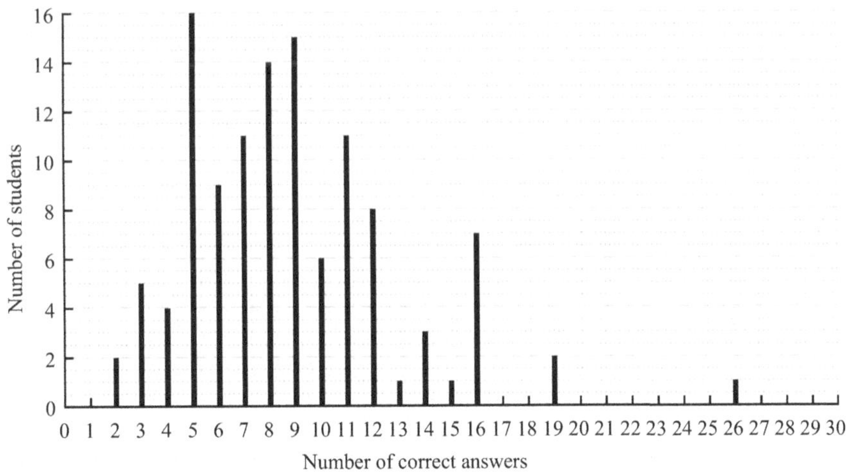

Fig. 1 Distribution of correct answers reported as scores

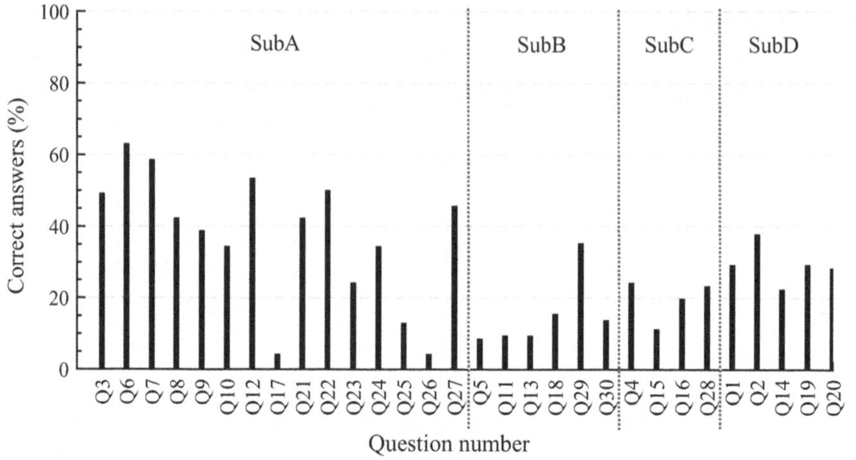

Fig. 2 Difficulty index of the different questions. The questions are grouped according to the categorization above described

where m is the number of possible student answers, n_i is the number of students who selected the choice i, and N is the total number of answers.

C_j takes a value in the interval 0–1. The higher the values, the more concentrated the answers. A value equal to 1 means that all students choose the same option, while $C_j = 0$ means that all the options have the same number of choices.

Figure 3 shows the C_j values for the 30 questions grouped according to the definition of the different subtests.

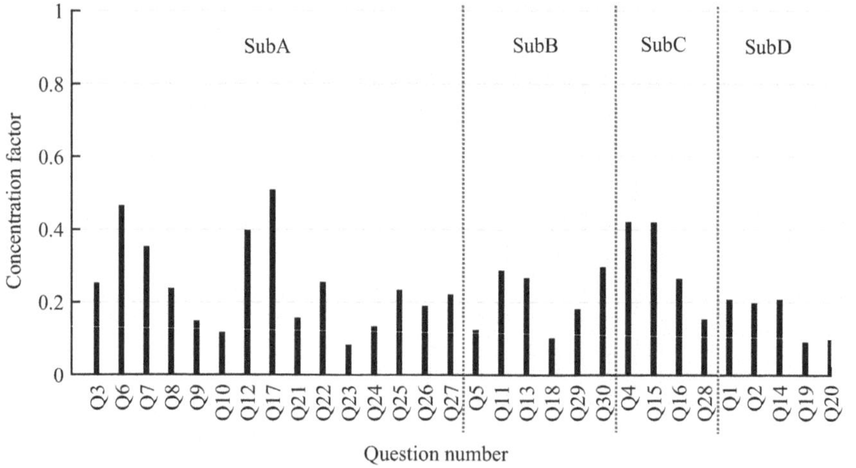

Fig. 3 Concentration factor of the different questions. The questions are grouped according to the categorization above described

Figure 2 shows a distinct gap in difficulty among many questions and subtests. Moreover, Fig. 3 makes evident how much the students' behavior in answering the questions is different. In some questions, the student's answers are concentrated on one/few choices. In others, they are spread among the various options.

In particular, SubA shows high values of correct answers in all items except two (Q17 and Q26). Q17 shows a high value of the concentration factor (c17 = 0.51). We verified that 68% of student answers concentrate on choice A. The analysis of answers to Q26 shows that 33% of students do not give an answer, and 55% of student answers are distributed among the choices A and B. In both cases, such choices support relevant misconceptions, as reported in the literature (Hake 1998).

SubB shows very low percentages of correct answers, except for question Q29. (It requires identifying forces in a static situation whereas the other three questions involve more complex dynamical situations.) Moreover, questions Q5, Q18, and Q29 show low concentration factors indicating that our sample does not answer on the basis of a precise model or misconception.

SubC shows very low percentages of correct answers, yet their concentration factors are, on average, high. The analysis of the number of choices to the different distractors makes evident that our students concentrated their answers on those distractors involving the misconceptions well known in the literature (Bao et al. 2002).

Questions of SubD show similar difficulty indexes and concentration factors. The questions Q19 and Q20 show low concentration factors since student choices are widely spread.

This preliminary analysis shows that students perceive the questions requiring the ability to recognize forces in dynamical systems and the understanding of Newton's third law, much more difficult than questions requiring the first or second Newtonian laws, or kinematics.

From research into student learning of mechanics, we know that student responses to problems analyzing dynamics of systems in many physical contexts can be considered as the result of their ability to apply a small number of mental models (Bao and Redish 2006). FCI has been accurately designed in a way to include different alternative models as distracters. As a consequence, individual student responses should be concentrated on the options associated with those models who he/she owns. On the other hand, if a generic student has little knowledge of the subject, he/she may act as if he/she has no models at all, or as if he/she chooses from a wide variety of different models. Therefore, knowing the distribution of the students' responses among the distracters can yield information on patterns in their reasoning strategies. We think that such information can be supplied by analyzing the relationships among all the answering options of the different questions (by studying, for instance, the coherence/incoherence or the context-dependence among the several sections of the questionnaire). We obtain those relationships through the identification of groups of students that exhibit the same patterns of reasoning. For this purpose, we use cluster analysis as we show below.

3.2 Cluster Analysis

We applied the k-means algorithm to each subtest (each conceptual dimension of the force concept) to identify common reasoning strategies. Then, we studied the relationships between each pair of results. Here we only report the analysis of the relationship between SubA and SubB. The full study can be found in Fazio and Battaglia (2019).

We analyzed our student sample with respect to their answers to the 15 questions classified as SubA, and we obtained as the best solution the one made of three clusters as reported in Table 2.

It is possible to characterize our student sample through the prevalent behavior in each cluster (Battaglia et al. 2017, 2019b; Di Paola et al. 2016). By looking at the clustering solution, we can find one prevalent behavior for each cluster.

Students in *Cl1_SubA* correctly answer questions Q3, Q6, Q7, Q9, Q12, Q22, Q24. Students in *Cl1_SubA* correctly use I and II Newton's Law for qualitative prediction of trajectories.

Questions Q8, Q17, Q25, and Q26 show high percentages of incorrect choices indicating prevalent thinking of our students connected to some persistent naïve conceptions as that "*persistence of original impetus*", "*largest force determines motion*", or "*motion is possible when forces overcome resistance*" (Hestenes and Halloun 1995).

Students belonging to *Cl2_SubA* cluster supply correct answers mainly to items involving the description of the motion of dynamic systems or the prediction of trajectories (as in questions Q6, Q8, Q12, and Q21). They fail to give correct explanations of the kinematic variable characterizing the chosen trajectories. Moreover, they do not supply an answer to questions that mainly require the ability to explain the characteristics of motions by applying the First or Second Newton's law.

Table 2 Result obtained by k-means algorithm

Cluster	Cl1_SubA	Cl2_SubA	Cl3_SubA
Number of students	11	39	66
Prevalent behavior	Students correctly answer questions that require the use of I and II Newton's law Some naïve conceptions as "persistence of original impetus"	Students correct answers questions involving the description of motion of dynamic systems. They fail to give correct explanations of the kinematic variables	They correctly use I and II Newton's law for prediction of trajectories but some questions show high percentages of incorrect choices. Some misconceptions as "a force due to the motion"

Number of students belonging to each cluster in the case of SubA and synthetic description of students' behavior

Table 3 Result obtained by k-means algorithm

Cluster	Cl1_SubB	Cl2_SubB	Cl3_SubB
Number of students	28	27	61
Prevalent behavior	Hybrid concept of force. Force as interaction and "force due to the motion"	The same behavior of Cl3_SubB and not answering half of the subtest items	Only the gravity force and the force due to the motion. Almost completely ignored forces as interaction

Number of students belonging to each cluster in the case of SubB and synthetic description of students' behavior

Students belonging to *Cl3_SubA* cluster supply correct answers mainly to questions Q6, Q7, Q12, and Q21. They correctly use I and II Newton's Law for the prediction of trajectories although some questions show high percentages of incorrect choices. Students in *Cl3_SubA* show misconceptions involving the characteristics of forces as "*a force due to the motion*" or "*a force in the direction of motion*" or "*an impetus force which is acting on the object after the object is no longer in contact with the agent applying the impulse*".

As previously pointed out, questions in *SubB* show low percentages of correct answers as well as low concentration coefficients. It means that the students' choices are spread among the different options. By clustering, we obtained again one best solution made of three clusters as reported in Table 3.

Cluster *Cl1_SubB* groups 28 students, from one side, correctly identify forces acting on a body as the result of the interaction between the body and the environment. From the other side, they identify a kind of force (the "*force due to the motion*" or "*supplied by an hit*") that seems directly connected to the body velocity or some hit supplied to it. Students belonging to this cluster share a "hybrid" (Ding and Beichner 2009) or "synthetic" (Gilbert and Boulter 1998) conception of force. It includes a composite idea that unifies different features of the naïve ideas ("*obstacles exert no force*" or "*motion implies active force*" or "*velocity proportional to active force*") and the scientifically accepted one. They correctly answer question Q29 which describes a static situation.

Students belonging to cluster *Cl3_SubB* see as forces acting on a moving body only the gravity force and the force due to the motion and almost completely forget the interaction forces. Such students share a naïve idea of force, widely discussed in the literature (see Brookes and Etkina 2009 and included references) and connected with many other misconceptions.

Cl2_SubB students only differentiate by those belonging to *Cl3_subB* since they do not answer half of the subtest items.

The *CLA* results for each subtest can supply information on students' understanding of each dimension of the force concept. We here analyze what kind of forces are identified in the analysis of dynamical systems of *SubB* by students that can correctly describe and explain motions of objects in the contexts of *SubA*. Moreover, it is worth investigating if their lacking competence in finding correct explanations can be related to a particular force concept. For these reasons, we identified what kind of classification was assigned in *SubB* to the students grouped into three clusters on the basis of their answers to *SubA*.

To make clear the results, we only describe the behavior of students belonging to *Cl1_subA* and *Cl3_SubA* since they supply more significant response patterns than *Cl2_subA* including students that do not supply any answer to almost half of the involved items.

We found that students belonging to *Cl1_subA* cluster are all included in *Cl2_subB*, which groups students showing the mixed force concept, previously described. We can infer that such students make proper use of the interaction idea as the starting point for the perception of Newtonian meanings. However, they maintain many naïve conceptions that cause wrong explanations when the contexts, where systems are moving, strongly triggers to make explicit such misconceptions.

Cl3_subB cluster contains the majority of students (49) belonging to cluster *Cl3_subA*. *Cl1_subB* cluster contains only 9 students. The naïve force concept makes these students unable to explain the characteristics of the described motions. Sometimes, such students can describe trajectories of moving bodies, but their correct answers can be more easily connected with similar observation of their everyday life motions rather than with a correct application of Newtonian laws.

It is also interesting to note that no one student belonging to *Cl1_subA* or *Cl3_subA* is included in the cluster *Cl2_subB*. It means that the *Cl2_subA* cluster contains all the 27 students belonging to the cluster *Cl2_subB*. Such two clusters share students characterized by a high number of no answers.

4 Conclusions

Our analysis allowed us to identify groups of students with similar responses, showing how such groups are characterized by correct answers as well as by non-correct answers and highlighting student misconceptions/non-normative conceptions. We analyzed such patters within two of the detected dimensions of the force concept and in the relationships among such dimensions.

Our CLA method (k-means) takes into account normative and correct as well as non-normative and incorrect student answers. This allowed us to identify in detail the status of their knowledge. For example, how the wrong understanding of the force concept can influence the correct understanding of motion characteristics in some contexts, as well as the incorrect understanding of motion characteristics in a different situation. It is obtained by grouping students that mainly show the same responses for a given group of questions (subtest). In this way, we recognize dominant or prevalent behaviors and infer the characteristics of models used by students in different situations.

The majority of our students cannot supply a Newtonian explanation of such motions (by applying the first and second Newton's law). They show in some contexts a correct ability to describe trajectories but not to explain them. This ability can be attributed much more to their direct experience with such kinds of motion than to a correct analysis. Only a small group can supply correct explanations, but only in given situations.

The majority of our students identifies as forces acting on a moving body only the gravity force and kinds of forces due to the motion ("motion force" or "impetus force") and almost completely forget the reaction forces. Such students share a naïve conception of force, widely discussed in the literature (see Brookes and Etkina 2009 and included references) and connected with many other misconceptions ("obstacles exert no force" or "motion always implies active force" or "velocity proportional to active force"). In parallel to this naïve conception of force, only a small group of students is starting to develop the Newtonian concept of force as an interaction between objects and environments. These two completely different conceptions seem to coexist in their mind, as it has been pointed out in Brookes and Etkina (2009). However, such a mixed idea of force makes these students able to identify correct relationships between force and motion in some situations, showing a good understanding of first and second Newton's law.

The sample and the instrument represent two limits of the analysis here discussed. We have analyzed a limited data set in terms of numerosity. This choice was driven by our need to verify the inferences caused by the quantitative results through the analysis of response protocols of each student. We felt this was relevant in this first application of the k-means method. Thus, we advise that the results of this analysis should not be extensively generalized.

In terms of the instrument, we should like to outline that our study regards the use of FCI questionnaire as a diagnostic instrument but cannot be necessarily generalized to all CIs. However, we see ways in which ClA of CIs can be useful at the level of the classroom. Making CIs more useful diagnostic instruments will allow researchers to connect instruction with student conceptions on a more fine-grained level and to help them in developing interventions that target the ideas in classes. Specifically, we found that the concept of force (investigated as student ability to identify forces acting in statics and dynamical systems) plays a relevant role.

We are able to draw educational implications from our results. These mainly focus on the need for teachers and instructors to be aware that students struggle with the ideas of force and motion mainly derived from their difficulty in understanding the ontological status of the force concept in physics (Brookes and Etkina 2009; Jammer 1957).

In Brookes and Etkina (2009), it is analyzed how many student difficulties with force and motion concepts are primarily due to a combination of linguistic and ontological difficulties. That paper shows how part of student learning is an act of negotiation of meaning between the instructor and the student, physics language, and categorizations and student ideas. Such a negotiation could involve putting students in different contexts where they can remove the uncertainty of meaning from a force concept as a cause of changing motion of a particular object and a force concept as an effect or property of motion.

The contexts exemplified by FCI items can provide the students with examples to make evident the reasoning strategies used by physicists in the historical development of concepts like force and motion (Jammer 1957). An implementation of learning approaches to Newtonian mechanics focusing on the relevant physics concepts, on the understanding of student alternative conceptions and of the forms of reasoning, they apply in different contexts can help students to speedup their appropriation of the main aspects of this theory. Moreover, teaching methodologies based on inquiry, computer-aided modeling, and hands-on and minds-on laboratory could lead to a better understanding of the physical quantities involved in the mechanical phenomena, as it has been highlighted by other researches (Pizzolato et al. 2014; Persano Adorno et al. 2015; Tarantino et al. 2010; Battaglia et al. 2010, 2019a; Fazio et al. 2006; Battaglia and Fazio 2018).

References

Allen DN, Goldstein G (2013) Cluster analysis in neuropsychological research: 13 recent applications. Springer Science+Business Media, New York

Bao L, Redish EF (2001) Concentration analysis: a quantitative assessment of student states. Am J Phys 69:S45–S53

Bao L, Redish EF (2006) Model analysis: representing and assessing the dynamics of student learning. Phys Rev Spec Top Phys Educ Res 2:010103

Bao L, Hogg K, Zollman D (2002) Model analysis of fine structures of student models: an example with Newton's third law. Am J Phys 70:766–778

Battaglia OR, Fazio C (2018) 2D simulation of wave-particle coupling inspired by walking droplets. Eur J Phys 39:045710

Battaglia OR, Fazio C (2021) Freshman engineering' reasoning strategies when answering FCI questions: a case study, Chap 15. In: Sidharth BG, Murillo JC, Michelini M, Perea C (eds) Fundamental physics and physics education research. Springer Science and Business Media LLC

Battaglia OR, Fazio C, Guastella I, Sperandeo-Mineo RM (2010) An experiment on the velocity distribution of thermionic electrons. Am J Phys 78:1302–1308

Battaglia OR, Di Paola B, Fazio C (2017) A quantitative analysis of educational data through the comparison between hierarchical and not-hierarchical clustering. Euras J Math Sci Technol Educ 13:4491–4512

Battaglia OR, Agliolo Gallitto A, Fazio C (2019a) Computer simulations to approach surface tension by means of a simple mesoscopic mechanical model. Comput Appl Eng Educ 27(6)

Battaglia OR, Di Paola B, Fazio C (2019b) Unsupervised quantitative methods to analyze student reasoning lines: theoretical aspects and examples. Phys Rev Phys Educ Res 15(2):020112

Brewe E, Bruun J, Bearden IG (2016) Using module analysis for multiple choice responses: a new method applied to force concept inventory data. Phys Rev Phys Educ Res 12:020131

Brookes DT, Etkina E (2009) "Force", ontology, and language. Phys Rev Spec Top Phys Educ Res 5(1)

Brophy S, Klein S, Portsmore M, Rogers C (2008) Advancing engineering education in P-12 classrooms. J Eng Educ 97:369–387

Cowgill MC, Harvey RJ (1999) A genetic algorithm approach to cluster analysis. Comput Math Appl 37:99–108

Di Paola B, Battaglia OR, Fazio C (2016) Non-hierarchical clustering as a method to analyse an open-ended questionnaire on algebraic thinking. S Afr J Educ 36(1)

Ding L, Beichner R (2009) Approaches to data analysis of multiple-choice questions. Phys Rev Spec Top Phys Educ Res 5(2)

Ding L, Caballero MD (2014) Uncovering the hidden meaning of cross-curriculum comparison results on the force concept inventory. Phys Rev Spec Top Phys Educ Res 10(2):020125
Erceg N, Aviani I, Grlaš K, Karuza M, Mešić V (2019) Eur J Phys 40
Everitt BS, Landau S, Leese M, Stahl D (2011) Cluster analysis, 5th edn. Wiley, Chichester
Fazio C, Battaglia OR (2019) Conceptual understanding of Newtonian mechanics through cluster analysis of FCI student answers. Int J Sci Math Educ 17(8):1497–1517
Fazio C, Guastella I, Sperandeo-Mineo RM, Tarantino G (2006) Measuring longitudinal wave speed in solids: two methods and a half. Eur J Phys 27:687–701
Gilbert JK, Boulter CJ (1998) International handbook of science education. In: Frazer B, Tobin K (eds). Kluwer, Dordrecht
Hake RR (1998) Interactive engagement versus traditional methods: a six-thousand-student survey of mechanics test data for introductory physics course. Am J Phys 66:64–74
Halloun IB, Hestenes D (1985) Common sense concepts about motion. Am J Phys 53:1056–1065
Hestenes D, Halloun I (1995) Interpreting the force concept inventory: a response to March 1995 critique by Huffman and Heller. Phys Teach 33:502–506
Hestenes D, Wells M, Swackhammer G (1992) Force concept inventory. Phys Teach 30:141–151
Huffman D, Heller P (1995) What does the force concept inventory actually measure? Phys Teach 33:138–143
Jammer M (1957) Concepts of force. Harvard University Press, Cambridge, MA
Jorion N, Gane BD, James K, Schroeder L, DiBello LV, Pellegrino JW (2015) An analytic framework for evaluating the validity of concept inventory claims. J Eng Educ 104:454–496
Lasry N, Rosenfield S, Dedic H, Dahan A, Reshef O (2011) The puzzling reliability of the force concept inventory. Am J Phys 79:909–912
Mantegna RN (1999) Hierarchical structure in financial markets. Eur Phys J B 11:193–197
Morris GA, Harshman N, Branum-Martin L, Mazur E, Mzoughi T, Baker SD (2012) An item response curves analysis of the force concept inventory. Am J Phys 80:825–831
National Research Council (2012) A framework for K-12 science education: practices, crosscutting concepts, and core ideas. The National Academies Press, Washington, DC. https://doi.org/10.17226/13165
Ott J (1999) Analysis of human genetic linkage, 3rd edn. Baltimore, London
Persano Adorno D, Pizzolato N, Fazio C (2015) Elucidating the electron transport in semiconductors via Monte Carlo simulations: an inquiry-driven learning path for engineering undergraduates. Eur J Phys 36(5)
Pizzolato N, Fazio C, Battaglia OR (2014) Open inquiry-based learning experiences: a case study in the context of energy exchange by thermal radiation. Eur J Phys 35(1):15–24
Planinic M, Ivanjek L, Susac A (2010) Rasch model based analysis of the force concept inventory. Phys Rev Spec Top Phys Educ Res 6:010103
Prince M, Vigeant M, Nottis K (2012) Assessing the prevalence and persistence of engineering students' misconceptions in heat transfer. J Eng Educ 101:412–438
Scott TS, Schumayer D, Gray AR (2012) Exploratory factor analysis of a force concept inventory data set. Phys Rev Spec Top Phys Educ Res 8:020105
Steif PS, Dantzler JA (2005) A statics concept inventory: development and psychometric analysis. J Eng Educ 94:363–371
Steif PS, Hansen MA (2006) Comparisons between performances in a statics concept inventory and course examinations. Int J Eng Educ 22:1070–1076
Steif PS, Hansen MA (2007) New practices for administering and analyzing the results of concept inventories. J Eng Educ 96:205–212
Streveler RA, Litzinger TA, Miller RL, Steif PS (2008) Learning conceptual knowledge in the engineering sciences: overview and future research directions. J Eng Educ 97:279–294
Tarantino G, Fazio C, Sperandeo-Mineo RM (2010) A pedagogical flight simulator for longitudinal airplane flight. Comput Appl Eng Educ 18:144–156
Wang J, Bao L (2010) Analyzing force concept inventory with item response theory. Am J Phys 78:1064–1070

Active Learning

Active Dissemination—Over Three Decades of Faculty Development in Active Learning

David R. Sokoloff

Abstract The activity-based physics group (with principal members Priscilla Laws, Ronald Thornton and the author) is best known for a number of physics active learning curricular developments including *Real Time Physics*, *Interactive Lecture Demonstrations*, and *Workshop Physics*. Equally important for the advancement of physics education has been the series of faculty development programs (institutes, workshops and short courses) that the group has organized and presented during more than three decades. This paper describes the context of these programs and the characteristics that have made them successful and sustainable over the years.

Keywords Active learning · Dissemination · Workshops · Institutes · Short courses

1 Introduction

During summer, 1987, the *National Microcomputer-Based Laboratory (MBL) Institute for Teachers of Physics* was held at the Center for Science and Math Teaching at Tufts University[1] (see Fig. 1). With financial support from the U.S. Department of Education, 20 exemplary secondary physics teachers from across the U.S. who had previously demonstrated their innovation and leadership, came to the Boston area to learn about new computer-based laboratory tools (for Apple II computers) and a new hands-on, active learning curriculum for use with the tools (that would eventually evolve into *Real Time Physics*) (Sokoloff et al. 2007, 2011). This first faculty development workshop organized by the *Activity-Based Physics Group*[2] incorporated a

D. R. Sokoloff (✉)
Department of Physics, University of Oregon, Eugene, OR 97403, USA
e-mail: sokoloff@uoregon.edu

[1] National Microcomputer Based Laboratory (MBL) Institute for Teachers of Physics, summers 1987–88, sponsored by U.S Department of Education Secretary's Discretionary Fund.

[2] The Activity Based Physics Group was awarded the 2010 APS Excellence in Physics Education Award. See https://www.aps.org/programs/honors/prizes/prizerecipient.cfm?last_nm=Sokoloff&first_nm=David&year=2010.

B. Jarosievitz and C. Sükösd (eds.), *Teaching-Learning Contemporary Physics*, Challenges in Physics Education,
https://doi.org/10.1007/978-3-030-78720-2_14

Fig. 1 Participants and staff of the *National Microcomputer-Based Laboratory (MBL) Institute for Teachers of Physics*, summer, 1987 at the Center for Science and Math Teaching, Tufts University (Ronald Thornton and the author are left and right, respectively, in the front row)

number of design features that supported the success of active learning dissemination. Since 1987, the author has been an organizer and/or presenter at nearly 300 institutes, workshops and short courses attended by over 8500 college/university and secondary faculty, sponsored by the National Science Foundation, U.S. Department of Education, Abdus Salam International Center for Theoretical Physics (ICTP), International Society for Optical Engineering (SPIE), UNESCO, Howard Hughes Medical Institute, and others. Because these programs have retained and refined these design features over the years, they have been successful in disseminating new developments in active learning. This paper will elaborate on this successful model for dissemination of active learning strategies, tools and curricula.

2 National Microcomputer-Based Laboratory (MBL) Institute for Teachers of Physics (1987)

The principal objectives of this institute were (1) to introduce the new computer-based tools and new active learning curricula through active participation of the workshop participants, (2) to help the participants clear up any gaps in their conceptual understanding of physics, (3) to provide resources and support for implementation at their institutions, and (4) to encourage dissemination to colleagues.

These design features were implemented through the following components found in almost all of our faculty development programs.

1. Participants read selections from the PER and active learning literature before coming to the workshop and are presented an introduction before beginning their work.
2. They take pre- and post-conceptual evaluation tests (which they will also use for action research with their students).
3. They work through the student curricular materials. (No lecturing on content!)
4. They prepare curriculum projects (plans) for implementation back home.
5. They receive sets of curricula and equipment to take home, and also information on procuring additional sets.
6. They receive modest monetary stipends for presenting active learning workshops to colleagues.

The advantages of this model for the participants are:

1. At the time of this institute, many teachers were not yet familiar with the burgeoning physics education research (PER) literature. This first component was designed to bring them up to speed.
2. Since testing is an important measure of the efficacy of any new curricular materials, it was important to introduce conceptual testing and action research. Since even the best teachers have some conceptual difficulties, the testing also served to make the participants aware of these without any stigma attached.
3. They were familiarized with the tools and curricular materials. Working through the materials is a stigma-free strategy for correcting any gaps in the teachers' understanding.
4. One of the most serious problems for teachers is finding time to assimilate new materials and strategies and adapt them to their teaching needs. Therefore, it was important that they left the workshop having developed a detailed plan for implementation.
5. Many teachers do not have the resources or the time to gather new equipment and curricula. Therefore, we assured that they left the workshop with the curricula and tools they needed for implementation.
6. While many of these teachers had already demonstrated leadership skills, we felt that incentives for "spreading the word" would make further dissemination more likely.

We were gratified to experience the interest of the participants in the then new computer-based data acquisition tools. What we did not anticipate was the difficulty of convincing them to implement the *Tools for Scientific Thinking* (Sokoloff and Thornton 1992) curricular materials that we had developed in mechanics, heat and temperature and sound. As we considered recruitment for the second summer institute of this program, we decided to invite the same group of participants back for a more in-depth exposure to active learning curricular reform. In fact, the inclusion of curriculum projects during the second summer was highly successful. Not only did we receive reports of successful implementations of the curricula, but we were pleased after that to see members of this group time and time again presenting workshops at national AAPT meetings.

3 Other National Summer Institutes for College and Secondary Faculty

Over the period 1990–2008, we were fortunate to receive funding for multiple series of summer institutes. Table 1 lists these. After *Real Time Physics (RTP)* (Sokoloff et al. 2007), *Workshop Physics (WP)* (Laws 2004), and *Interactive Lecture Demonstrations (ILDs)* (Sokoloff and Thornton 2004) were published, these were the main active learning curricula explored in these institutes. In the later years, new topics such as *Physics with Video Analysis* (Laws et al. 2009), the use of personal response systems, and video with *ILDs* and collaborative problem-solving tutorials (Johnson 2001) were introduced. In all of these institutes, participants were either provided with equipment, or their schools or school districts were required to commit to purchasing equipment as a prerequisite for attendance. Of course, participants were provided with one set of all printed materials included in the institutes. Figure 2 shows a group of college faculty with Priscilla Laws at one of the *Summer Seminars* at Dickinson College (before the rooms at Dickinson were re-designed to enhance the implementation of *Workshop Physics).*

Are institutes of this type effective? According to the independent evaluator's *final report* for the Activity-Based Physics Faculty Institutes (ABPFI)— the last set

Table 1 Active learning summer institutes 1990–2008

Institute name	Years	Sponsor	Participants
Summer Seminars: Teaching Introductory Physics Using Interactive Teaching Methods and Computers	1990–97, Dickinson 1998, Oregon	NSF	College/university
Student Oriented Science (SOS): Microcomputer-Based Labs for Secondary Science Teaching	1994–96, Tufts, Oregon, Dickinson	DOE	High school
Activity-Based Physics Institutes (ABPI)	2000–03, Dickinson, Oregon	NSF	High school
Activity-Based Physics Faculty Institutes (ABPFI)	2005, 2007, Oregon 2006, 2008, Dickinson	NSF	College/university/high school

Fig. 2 Participants with Priscilla Laws at one of the Dickinson College Summer Seminars (before the room was remodeled to better support *workshop physics)*

of nationwide institutes we presented under NSF support—"… given the extremely positive feedback participants gave regarding the workshops and their post-ABPFI experiences with activity-based teaching, combined with the number of faculty trained over the course of four years, it would be hard to describe the ABPFI as anything but extremely successful overall. By the end of the grant period, 170 college instructors had attended an ABPFI workshop. As a result, nearly 20,000 students are now impacted by ABPFI trained faculty each year, including over 5000 minority students and nearly 9000 women. All the peer-reviewed quantitative research on the impact of activity-based learning in physics education corroborates the qualitative reports that ABPFI trained faculty almost universally give: activity-based pedagogy leads to substantial gains in student learning. That thousands of students, many of them members of traditionally underrepresented populations, are able to take advantage of these learning gains every year as a result of ABPFI is a testament to the ABP group whose members so effectively executed this grant" (The Activity-Based Physics Faculty Institutes 2010).

4 National Chautauqua Short Courses

July, 2019 was the 25th anniversary of the first 2.5-day, Chautauqua short course on active learning that we presented. During those 25 years, we have presented 37 of these courses to over 1000 introductory physics faculty.[3] These were mostly held

[3] See https://pages.uoregon.edu/sokoloff/CHAUT.htm.

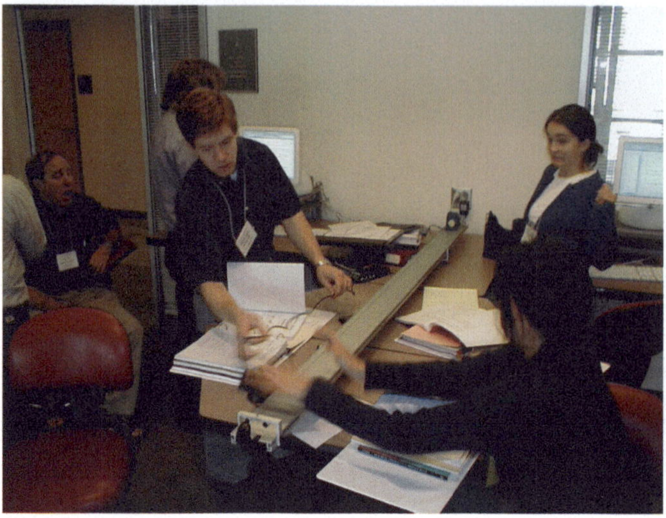

Fig. 3 Participants with the author at a Chautauqua short course, held at Dickinson College

at University of Oregon, Vernier Software and Technology and Dickinson College, although we also presented several over the years in Australia, Puerto Rico and Hawaii. Figure 3 shows the author and participants at one of these Chautauqua courses.

While this national program was initially intended for college-level STEM faculty and sponsored by a large NSF grant administered through University of Pittsburgh and the California State University system, in recent years, our offerings have been self-supporting, and a significant contingent of high school teachers have attended each year. One factor that has made these courses more attractive to high school teachers is that since 2011, teachers have had the option of earning up to three graduate credit hours through the University of Oregon Physics Department, by attending the course and completing curriculum projects as assignments afterward.

5 Active Learning in Optics and Photonics (ALOP)

Beginning in 2003, Dr. Minella Alarcon, program specialist (now retired) for physics and mathematics at UNESCO, Paris, worked with an international team of physics educators to develop a five-day workshop for teachers of introductory physics called active learning in optics and photonics (ALOP).[4] UNESCO originally coordinated and provided basic funding for the project, but more recently support has come principally from the International Society for Optical Engineering (SPIE) and the Abdus Salam International Center for Theoretical Physics (ICTP). The pedagogy of the

[4] See https://pages.uoregon.edu/sokoloff/ALOPwebpage.html.

ALOP curriculum and the structure of these workshops is based on the same principles used in the active learning short course (Chautauqua) presented in Melbourne, Australia in 1999, which Dr. Alarcon attended.

ALOP workshops:

1. Are designed for secondary and first year college faculty in developing countries,
2. Include teacher updating and introduction to active learning approaches,
3. Are locally organized,
4. Use simple, inexpensive apparatus available locally or easily fabricated,
5. Are designed and presented by an international team of volunteer teacher educators,
6. Provide a PER-based conceptual evaluation *(LOCE)* to measure student learning,
7. Provide complete teachers' guides, and
8. Distribute equipment sets to facilitate local implementation.

Figure 4 shows the cover of the *ALOP Training Manual* (Sokoloff (ed) 2006), published by UNESCO and edited by the author. The original ALOP team (now led by Dr. Joseph Niemela of ICTP) includes Zohra Ben Lakhdar (University of Tunis, Tunisia), Vasudevan "Vengu" Lakshminarayanan (University of Waterloo, Canada), Ivan Culaba and Joel Maquiling (Ateneo de Manila University, The Philippines), Alex Mazzolini (Swinburne University of Technology, Australia), and the author.[5] More recent but significant additions to the team have included Souad Lahmar (University of Tunis, Tunisia), Khalid Berrada (Cadi Ayyad University, Morocco), Cesar Mora (National Polytechnic Institute, Mexico), and Angela Guzman (University of Central Florida, U.S.).

Figure 5 shows the geographical distribution of the 37 ALOP workshops presented during 2004–2019 to over 1100 faculty. There has also been a second generation of ALOPs taught by previous participants in their local languages. The ALOP Training Manual has been officially translated into French and Spanish. There are also unofficial translations in Arabic, Armenian, Thai, Ukrainian, and Georgian. Figure 6 shows the author and participants in the October, 2019 ALOP workshop in Bandung, Indonesia.

ALOP's intensive workshop illustrates active learning pedagogy through carefully crafted learning sequences that integrate conceptual questions and hands-on activities like those found in *RTP* and *ILDs*. ALOP has adapted most of the design features mentioned above. Of course, there are some limitations. For example, complete equipment sets are distributed only to roughly one third of the participants, making sure that at least one set is available in the region of each participant. The equipment and materials are generally inexpensive enough that participants can easily gather duplicate sets. While there are no planning assignments within the workshop, time is allotted to discussions of implementation issues.

[5] The ALOP team was awarded the 2011 SPIE Educator Award. See https://spie.org/about-spie/awards-programs/awards-listing/spie-maria-j-yzuel-educator-award.

Fig. 4 ALOP training manual, published by UNESCO

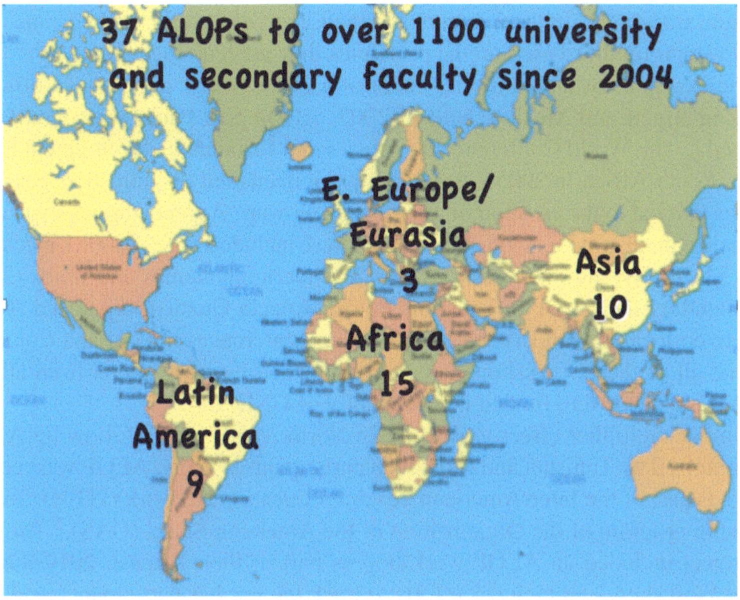

Fig. 5 Geographical distribution of ALOPs presented since 2004

Fig. 6 Author with participants in an October 2019 ALOP workshop in Bandung, Indonesia

The *ALOP Training Manual* includes five modules: (1) Introduction to Geometrical Optics, (2) Lenses and Optics of the Eye, (3) Interference and Diffraction, (4) Atmospheric Optics, (5A/B) Optical Data Transmission, and Wavelength Division Multiplexing. Each of these has embedded practical applications that are designed to intrigue students, help them understand their everyday world, and become aware of career opportunities based on the principles they are learning.

Some examples of the multiplier effect of a single ALOP are the spin offs from ALOPs in Brazil and Mexico, held in 2007. ALOP Sao Paulo was one of three mandated by the World Conference on physics and sustainable development held in Durban, South Africa in 2005.[6] Argentine participants led by Julio Benegas began meetings that led to the creation of a series of followup "Aprendizaje Activo" workshops held annually in La Falda, Argentina. These included optics (2008), mechanics (2009), electricity and magnetism (2010), and thermodynamics and fluids (2011). After attending the ALOP in San Luis Potosi, Angela Guzman organized ALOPs that were co-located with LACCEI engineering conferences in Peru and Colombia. Note that all of these workshops have been attended by participants from all over Latin America and were funded independently of UNESCO.

A similar multiplier effect took place in North Africa where follow up ALOPs were organized by Tunisian and Moroccan participants. Finally, ALOP was recently adopted as part of the Inter-American Teacher Education Network (ITEN), Teacher Fellowship program of the Organization of the American States (OAS).[7] This two-year project included an ALOP workshop as part of their August, 2019 Summer Teacher Workshop, and includes efforts to track the teacher participants, and study their future use of active learning strategies in their teaching, and the impact of these strategies on their students.

Acknowledgements The author thanks the members of the Activity-Based Physics Group, especially Priscilla Laws and Ron Thornton, for their assistance in developing and presenting faculty institutes, workshops and short courses over the years. He especially thanks the secondary and college/university faculty who have participated in these programs, contributed suggestions for their improvement and used the knowledge they gathered to improve the teaching and learning of physics.

References

Johnson M (2001) Facilitating high quality student practice in introductory physics. Am J Phys 69(S1):S2. https://doi.org/10.1119/1.1286094

Laws P (2004) The workshop physics activity guide, modules 1–4. Wiley, Hoboken, NJ

Laws P, Teese R, Willis M, Cooney P (2009) Physics with video analysis. Vernier Software, Portland, OR

Sokoloff D (ed) (2006) Active learning in optics and photonics training manual. UNESCO, Paris

Sokoloff D, Thornton R (1992) Tools for scientific thinking—motion and force and heat and temperature, curricula and teachers' guides, 2nd edn. Vernier Software, Portland, OR

Sokoloff D, Thornton R (2004) Interactive lecture demonstrations. Wiley, Hoboken, NJ

Sokoloff D, Thornton R, Laws P (2007) Real Time physics: active learning labs transforming the introductory laboratory. Eur J Phys 28:S83–S94. https://doi.org/10.1088/0143-0807/28/3/S08

---

[6] See https://www.aps.org/publications/apsnews/200512/conference.cfm.

[7] See http://www.oas.org/en/iten/default.asp?pagename=fellowship.

Sokoloff D, Thornton R, Laws P (2011) Real Time physics: active learning laboratories, 3rd edn, module 1: mechanics, module 2: heat and thermodynamics, module 3: electricity and magnetism and module 4: light and optics. Wiley, Hoboken, NJ

The Activity Based Physics Faculty Institutes (2010) From final report for national science foundation, course, curriculum and laboratory improvement national dissemination grant (CCLI-ND)

Strategies for Active Learning to Improve Student Learning and Attitudes Towards Physics

Claudio Fazio, Marina Carpineti, Sergej Faletič, Marco Giliberti, Gareth Jones, Eilish Mcloughlin, Gorazd Planinšič, and Onofrio Rosario Battaglia

Abstract Over the last several years, active learning methods and strategies have received considerable attention from the educational community and are commonly presented in the related literature as a credible solution to the reported lack of efficacy of more "traditional" educative approaches. Research has shown that a possible factor is the strongly contextualized nature of active learning that focuses on the interdependence of situation and cognition. In this paper, we report the results of a Symposium with different contributions in the field of research on active learning. We start with a system analysis of the mental processes involved in learning physics which explains how active learning involves cognition followed by a response and feedback. Then, we describe a novel approach to active learning in which students participate in

C. Fazio (✉) · O. R. Battaglia
Dipartimento di Fisica e Chimica, Università degli Studi di Palermo, Palermo, Italy
e-mail: claudio.fazio@unipa.it

O. R. Battaglia
e-mail: onofriorosario.battaglia@unipa.it

S. Faletič · G. Planinšič
Faculty of Mathematics and Physics, University of Ljubljana, Ljubljana, Slovenia
e-mail: sergej.faletic@fmf.uni-lj.si

G. Planinšič
e-mail: gorazd.planinsic@fmf.uni-lj.si

M. Carpineti · M. Giliberti
Dipartimento di Fisica, Università degli Studi di Milano, Milano, Italy
e-mail: marina.carpineti@unimi.it

M. Giliberti
e-mail: marco.giliberti@unimi.it

G. Jones
Department of Physics, Imperial College London, London, UK
e-mail: w.g.jones@imperial.ac.uk

E. Mcloughlin
School of Physical Sciences, CASTeL, Dublin City University, Dublin, Ireland
e-mail: eilish.mcloughlin@dcu.ie

B. Jarosievitz and C. Sükösd (eds.), *Teaching-Learning Contemporary Physics*, Challenges in Physics Education,
https://doi.org/10.1007/978-3-030-78720-2_15

theatrical activities involving physics topics. Following this, we present experience of implementation of active engagement methods used in physics teacher preparation courses. Finally, a study is reported on an inquiry-based learning approach carried out in the context of socio-scientific issues with pre-service physics teachers.

Keywords Active learning · System analysis of mental processes · Physics theatre in learning · ISLE framework · Pre-service science teacher inquiry

1 Introduction

There is today a wide consensus that to improve student learning the traditional lecture format, where students passively receive information, should be evolved towards an approach inducing specific student engagement and activity in learning. In the "active learning" approach, students do more than just listen to a lesson. They are engaged in actively reading, writing, posing and discussing questions, gathering data from different sources, building models and in solving problems ultimately aimed at developing their knowledge, skills and attitudes. In active learning, students are involved "in doing things and thinking about the things they are doing" (Bonwell and Eison 1991).

Active learning methods and strategies are credited with improving student conceptual understanding in many fields, including physics. Research has shown that improvement arises from the strongly contextualized nature of education that active learning brings that focuses on the interdependence of situation and cognition. When learning and context are put together, knowledge is seen by learners as a tool to be used dynamically to solve problems and to ultimately develop critical transversal skills, rather than knowledge being the final product of education.

For these reasons, active learning has gained strong support from teachers and faculties looking for effective alternatives to traditional teaching methods. Moreover, active learning activities developed in the context of socio-scientific issues (SSI), which are contemporary and relevant scientific topics with moral or economic implications are today widely promoted to enhance student's scientific literacy. SSIs centre around a range of scientific, social or moral viewpoints may conflict with the students' own views and thus makes them personally relevant to students.

In this paper, we present the contributions to a Symposium organized by the GIREP thematic group on active learning strategies, starting with a systems analysis of the basic mental processes involved in learning physics. This analysis helps to clarify the key causal links and feedback processes in learning which lead to advantages and disadvantages of various forms of active learning. Following this, a novel approach to active learning is presented, based on the acknowledgement that scientific theatre can be an extremely useful tool to actively involve students, stimulating motivation and arousing positive emotions. Then, active engagement methods used in high school physics teacher preparation courses and their implementation are described. For the implementation, it is crucial to find a well-researched and tested active engagement approach, and to this aim, the well-known investigative science learning environment approach is chosen as a theoretical framework, as it clearly emphasizes experiments and observations, which are well acknowledged as crucial for the epistemology of physics. Finally, a study reporting on an inquiry-based learning approach carried out in context of socio-scientific issues with a cohort of pre-service physics teachers is discussed. This study aimed at developing the pre-service teacher's knowledge and skills for adopting inquiry and SSI contexts in their classroom practices.

2 A Systems Analysis of the Mental Processes Involved in Active Learning (Based on the Presentation of Gareth Jones at the Symposium)

There are various forms of active learning (Fazio 2020); but in all cases, their efficacy lies in the individual student's mind being active in producing a response on a short time scale to an intellectual stimulus. Thus, an analysis of the mental processes involved in learning can be expected to lead to a better understanding of the advantages and disadvantages of particular learning and teaching methodologies and hence their improvement. This is particularly true of higher education in physics where the aims include the achievement of a deep understanding of cognitively challenging concepts and methods which go far beyond the remembering of large amount of factual information and the acquisition of skills. In physics, thinking is much more important than remembering. Also, the importance of considering mental processes is related to the fact that the essence of learning physics involves each individual asking her/himself questions such as "Why? How? Do I understand?" This implies that learning physics is mostly internally focussed and involves cognitive loops. Of course, self-posed questions can also be posed to others. When others are involved in discussion or Q&A, e.g. in many forms of active learning, the key learning process is LISTEN → THINK → RESPOND.

To analyse this further takes us into the realms of cognitive psychology and cognitive neuroscience which are large research fields in their own right where important developments are occurring but whose applications to the practice of physics higher education are limited although of increasing interest (Redish 2014). The difficulty lies in the extreme complexity of networks in the brain involving very large numbers of neurons with varying connectivity. Also, the dynamics of the connectivity appear to be crucial to the learning process, and this is very difficult to model. A possible way forward would be to attempt a systems analysis of the mental processes involved in active learning which is based on (or at least is consistent with) our present knowledge of basic neuroscience. Systems analysis is best developed by means of "flow diagrams" which clarify the causal links, and the operation of various loops and feedback mechanisms and other processes involved in learning even though they cannot easily represent changes in connectivity. Such flow diagrams are analogous to the use in electronics of circuit diagrams to describe processes which ultimately are based on Maxwell's equations and atomic physics.

2.1 A Particular Model of Mental Processes Involved in Learning by an Individual Student

An example of a flow diagram is shown in Fig. 1. This is a simplified representation of the mental processes involved in learning by an individual student and has a pragmatic purpose of aiming to help the design of learning events. It aims to be consistent with existing cognitive neuroscience but not embedded in it since it attempts to incorporate experience of producing flow diagrams produced for Monte-Carlo modelling of processes in physics experiments. Although it is not particularly adapted to active learning, the intention is to demonstrate its efficacy from general considerations. The "boxes" are meant to represent operational processes, while the connecting lines represent causal links and information flows. It starts on the extreme left with a box labelled "Learning Event" which could be a lecture, tutorial, reading, laboratory experiment, etc. This could be broken down into individual events whose duration and information content could be quite small. Each is perceived by the student's sensory systems "Look", "Listen", etc. "Read" is included as distinct from "Look" since it involves a particular form of semi-automatic decoding of sensory input from the eyes or from the sense of touch (e.g. in braille). Each of these immediate sensory information flows is then processed or decoded (at different speeds) in

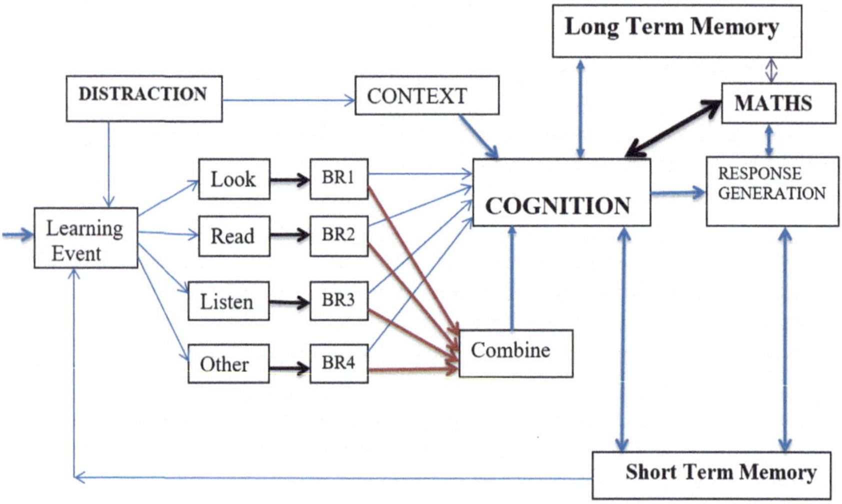

Fig. 1 Basic mental processes involved in learning physics. Note the various feedback loops and the specific MATHS loops. BR refers to specific regions of the brain which process sensory signals

particular brain-regions (labelled "BR1", etc.), and the outputs are then combined or integrated as well as being sent directly to a box labelled "cognition". The "Combine" box represents a pre-cognition form of integration of inputs before the main thinking occurs in the "Cognition" box. The "Cognition" box is meant to represent the processes occurring in the brain involving interpretation, logical deductions, thinking emotionally (like/dislike), formulation of ideas, speculation, etc. It is affected not just by sensory inputs, but also by thoughts related to prior knowledge, the context of the specific learning event as well as inputs from long-term memory enabling comparisons with previous knowledge or assumptions to be made. The results of cognition will be stored in short-term memory. It must be emphasized that links of all the boxes to both short-term and long-term memory are ubiquitous and to show them all in Fig. 1 would create confusion. Clearly, there are other ways of constructing such flow diagrams.

Also shown in Fig. 1 is a box labelled "Distractions" which is shown as an input to the learning event and also indirectly to cognition. The important point is that there could be extraneous sensory inputs which will also be processed. They could be organized and difficult to separate from the genuine learning event, e.g. another student talking. The filtering out of such inputs is a contribution to cognitive load which depends on the context and the student's prior knowledge. Consideration of prior knowledge is helpful in learning physics both through comparison with new knowledge and also because it can lead to the questioning of assumptions and hence improved understanding. This is difficult to represent in a systems diagram, and so, prior knowledge is shown as a separate box even though it is a part of long-term memory. The way that both prior knowledge and distraction affect cognition depends on the particular context of the learning event, and so, the diagram shows them linked to the cognition box via a context box. It should be noted that this diagram should be regarded as a pragmatic aid to designing and analysing learning processes and to act as a starting point for computer simulation.

A characteristic of passive learning of factual information is that the cognition may be minimal and will be followed by a loop to the next individual learning event, e.g. the next equation or sentence or paragraph in a book. In active learning, there should be a "Response Generation" which could take several forms, which could be expressed by writing or speaking or some other physical response (think of Archimedes jumping out of his bath and exclaiming "eureka"!). The response box is at the heart of the gains which come from active learning and can sometimes be separated into the products of fast and slow cognition involving several iterative cognitive processes. The aphorism "engage brain before speaking" is apposite here. This response would be stored in short-term memory and after more cognition transferred to long-term memory. In most of physics, the use of maths is a major component of learning, and this is shown in the diagram. Such maths may be very extensive and is the mechanism of logical deduction and clarification of the cognitive process. Feedback via loops involving response generation will refine and advance the results of cognition.

Not shown in the diagram are extra features to represent processes which deal with cases where there is a blockage caused by lack of understanding of the main sensory inputs or the results of cognition. When this happens, the student needs to use library resources or an "explanation" from an instructor who has a good understanding of both the subject and the student's difficulty. In laboratory work, it can be the start of a new investigation involving measurements. There is extensive relevant literature in cognitive science which underlies the above section, and an example is Chi et al. (2018).

2.2 An Example Illustrating Interaction with Other Students

The same methods can be extended to produce diagrams which include interactions with the minds of other individuals such as other learners and/or an instructor. To make such diagrams more useful, they need to be simplified by omitting some links, e.g. to memory. Since there are many forms of active learning, many such diagrams could be designed to analyse the processes involved with the aim of improving our understanding of their advantages and disadvantages. For example, they make clear the importance of careful responses and the pointing out of faulty reasoning which can then be corrected. An example is the upward spiral of "Socratic Dialogue" teaching which can occur in tutorials and where feedback loops can be very effective.

They also help us to understand the process difficulties in group work encountered by particular students, e.g. with introverted personality. An example could be the difficulties experienced in some forms of active learning by a student who has "High Functioning Autism" (or more generally ASD, Autism Spectrum Disorder). Such students are fairly common in university physics classes, but the condition is often undiagnosed. They have a strong internal focus and difficulty in interacting with others. But this is often compensated by high ability in deep thinking, great accuracy and care over detail, and creative thinking. They are sometimes referred to as "Twice Exceptional" students to recognize that their creativity and care over detail makes them exceptional also in a positive way.

As an example, Fig. 2 represents "Learning by Student Discussion Groups" following a prior reading assignment. These often involve students explaining or sharing knowledge and understanding. Such methodologies have the advantage of requiring a clear shift from passive to active by requiring short-term responses, critical thinking and responses from students followed by rapid feedback from others. But such active learning methodologies have the potential disadvantage that some students may be misled or confused by interactions with persuasive peers who may have limited understanding and/or have erroneous views. This may cause real learning problems in students who are not confident of their own knowledge and understanding.

The loops involved consist of one student "i", speaking followed by other students (shown as Student "j") listening. Student "j" may think based on her/his prior knowledge and then in turn will speak. If student "j" has ASD, then the thinking may be quite profound, but she/he will find it difficult to form a response and may produce a delayed response or no response at all. This is shown in Fig. 2. The loop consists of other students listening and thinking about the response, but they may do little

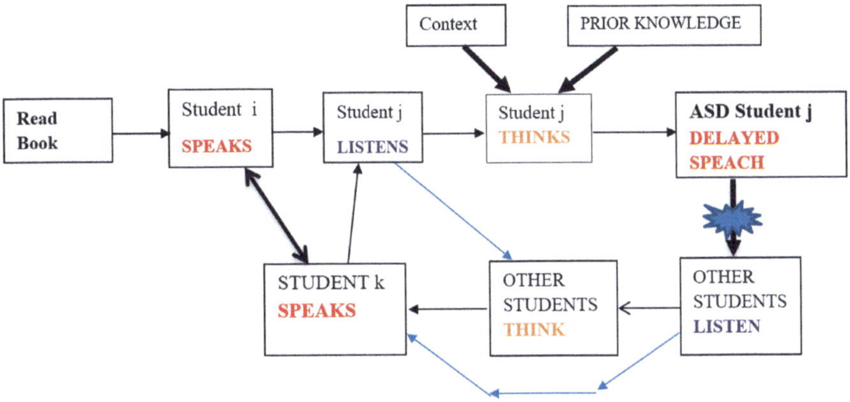

Fig. 2 Learning by peer discussion. Note that a student with ASD may be excluded

listening and thinking and so may bypass the Think box. The way the loops may then operate is that the discussion may be dominated by just two students ("i" and "k") who may just blurt out their opinions to impress others by being quick in responding. But for introverted students, quiet internal contemplation and precise rational thought processes can be more productive for them than discussions with extrovert peers whose pronouncements they find very difficult to challenge. So, this internal approach to learning should not be discouraged by the teaching methodology employed. Quiet contemplation should be allowed, indeed encouraged. This illustrates one of the dangers of such student discussion groups. Most published accounts of active learning ignore the effect of the range of student personality types. But students vary and ASD is just one example where variation makes a difference (Andersen 1993).

2.3 *Special Features of Advanced Physics Relevant to Active Learning and the Above Diagrams*

There are some particular features of physics which make it suited to such an analysis as described above. Physics is devoted at heart to achieving deep fundamental knowledge and understanding (at an internal personal level) of the physical world and the universe as a whole; indeed, there is substantial evidence that this feature is very important in inspiring young people to study physics at university. Physics also involves going far beyond what can be directly observed (or inferred) and requires ways of thinking that question assumptions and challenge what seems obvious. The worlds of the very small and the very large are different to the world we live in so mental processes that have evolved in response to the need to succeed in everyday activities in our world may not be the most effective for learning physics; the brain is a product of evolution. Challenging questions that students may have should be addressed in discussions with experienced physicists. Finally, it should be noted that the general method of a systems-based approach to modelling mental processes in learning is similar to the detailed simulation of physical processes involved in some experiments (e.g. experimental particle physics) using Monte-Carlo methods (e.g. Akerib et al. 2012) in testing our understanding of the basic physics involved. It is a kind of "imitation game", to borrow a phrase first used by Turing in a more profound but related context (Turing 1950). A long-term ambition could be to use Monte-Carlo methods to explore models of learning.

3 Physics Theatre to Foster Active Learning in Students (Based on the Presentation of Marco Gilberti at the Symposium)

Getting scientific notions perceived as beautiful, interesting and useful are a fundamental aspect to take into account to generate interest and curiosity in people towards science (Carpineti et al. 2006). Although the presence of different stimuli (auditory, visual, of participation, of discussion…) added together in a multimodal learning structure determines the growth and the internalization of concepts, especially the scientific ones, at school we often tend to make a prevalent use of the auditory stimulus (Carpineti et al. 2016). Student listens to the teacher's explanation with only few interventions of other factors, so that the relationship between students and teacher easily becomes authoritarian, while the dominant language is verbal, descriptive and procedural.[1] This is particular true for the hard sciences, such as physics, that students often perceive as having only one "face" with little or no room for creativity. In order not to neglect the other faces of hard sciences, an "A", standing for Art, has been added to the STEM disciplines to propose an integrated STEAM approach to sciences (EPS Physics Education Division 2012).

In this perspective, scientific theatre deserves particular attention since it is an extremely useful tool to stimulate motivation from positive emotions (Carpineti et al. 2006, 2016; Barbieri et al. 2015). That is the reason why, in the last 15 years, the theatrical approach to physical themes has been explored by the group "Lo Spettacolo della Fisica"[2] founded by M. Carpineti, M. Giliberti and N. Ludwig of the Physics Department of the University of Milan.

The educational use of scientific theatre at school, with the direct involvement of students, can be proposed in various forms; typically, through improvisation, role-play or writing and recitation of a script. In the first case, students impersonate a role without a prior preparation and outline aspects and problems concerning a given scientific topic. In the role-play case, on the contrary, students impersonate specific roles after having reflected and studied in groups, at school and/or at home,

[1] TEMI 2016 Homepage. https://cordis.europa.eu/project/id/321403, last accessed 2016/11/21.

[2] Lo spettacolo della fisica Homepage, http://spettacolo.fisica.unimi.it, last accessed 2021/01/13.

reporting debates and controversies concerning scientific issues, often with social-ethical repercussions. Finally and generally, the recitation of a script written by students focuses, instead, on historical or particular facts concerning a scientific event. Being much more challenging, this last activity gives students more freedom of expression and allows the development of more creativity.

There is, however, another common approach to the use of theatre that starts from the vision of shows held by professional actors. The theatrical performance is often followed by a debate between actors and spectators on the covered topics, according to an idea that can be called that of "Theatre of Debate". The purpose is to bring people, mostly children and young people, towards the world of science both, by generating enthusiasm and interest for the wonder of science and by considering the most debated social scientific issues of the moment. In particular, some theatre companies—not rarely also made up or with the advice of university physics professors—have become active by performing real physics experiments to show students the wonder of physics, or as a tool to discuss aspects that are often neglected at school and in textbooks, with new points of views to propose a change of perspective on the discipline.

Despite its value in teaching, the scientific theatre performed by theatre companies cannot be a didactic theatre. On the contrary, it must be a theatre for everyone, not for schools, at least in the same sense in which the "Hamlet" by Shakespeare is for everyone: it has not been written for school; but, for its beauty and for the ideas it expresses, it is studied by everyone right at school.

Professional scientific theatre, in short, does not have to talk about physics in order to explain it or to divulge it, but has to reflect on what, perhaps, we humans have of most important: our vision of the world and of life, of which they are part, together, the things we know, the ways we know them and the position and meaning we give ourselves in relation to the world. In dramas and comedies, there is a conflict that must be resolved by the interaction with the other characters. This fact keeps the viewer alert and active. If the conflict/game manifests itself in questions about physics, an emotional involvement may arise. This involvement often generates participation and, if opportunely stimulated, also an active interest towards physics itself.

3.1 Active Learning Starting from Scientific Theatre

While the desire to understand is in general natural, in most cases, the possibility to understand is not spontaneous, but is to be learned, so that knowledge and culture can be internalized in a personal way that leads to the appropriation of concepts. When this happens, the conceptual change occurred makes the person become able to process the knowledge received and return it to society. It is a matter of fact that schools are not, and cannot be, enough for this process to happen. In fact, culture is only partly scholastic and learning certainly does not end at school.

The physics education division of the European physical society has recently highlighted the importance of learning physics for the whole life, even in informal contexts (Carpineti et al. 2016). For this reason, schools have to open up and interact more and more with society. To help this interaction, formal, non-formal and informal education have to be in close synergy with each other, while the person must be at the centre of the teaching process. In fact, the learning of physics in a formal context (i.e. at school) depends very much on what is said, read, seen, experienced about physics in contexts that are not formal. And, on the other hand, the way we teach physics at school has great repercussions on the social and cultural image of the physics itself.

Theatre can be an important tool for generating school-society-person interactions. In fact, it is able to develop scientific imagination, to help improve learning by using emotions in an effective way, to promote an approach to physics through the channel of affectivity and to enhance personal needs, thus diminishing cultural and gender discrimination by promoting a scientific culture that is more profound and even more human.

3.2 Theatre to Bridge the Gap Between Informal and School Activities in Milan: An Example

To foster active learning in students, the work proposed by the physics education research group of the University of Milan aims at creating a bridge among informal, non-formal and school activities. As an example, we briefly describe the general structure of a path concerning the electromagnetic spectrum.

- The first step (vision of a show held by professional) is the vision of "Light from stars" a physics theatre show, offered by "Lo spettacolo della Fisica", written and performed by M. Carpineti, M. Giliberti and N. Ludwig (Fig. 3). In this show, the principal theme, that is the vision of the universe at various wavelengths, is

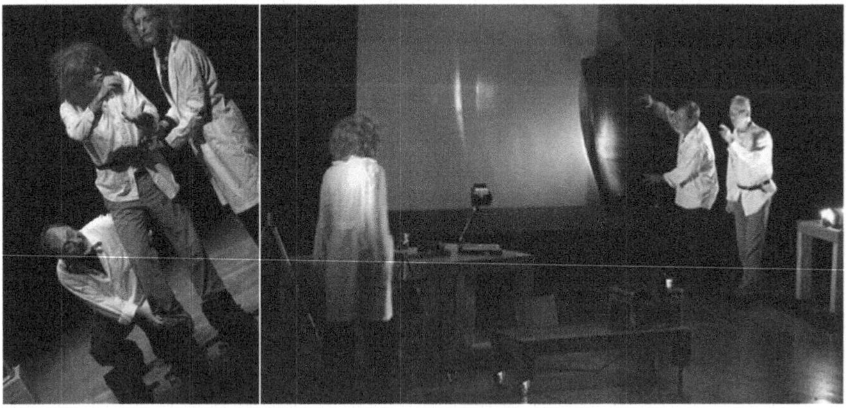

Fig. 3 Two pictures of the show "Light from stars". *Credits* Fabrizio Favale

embedded in those of the effectiveness of the scientific communication and of gender gap

- The second step is the participation of students and their teacher in the so-called PLS (Scientific Degree Plan) laboratory "The invisible colours at the edges of rainbow" (a non-formal activity) held at the physics department of the university of Milan. The laboratory (of about 3 h) proposes students to explore in small groups the electromagnetic spectrum at the borders of the visible with simple experiments using thermal cameras and UV cams connected to tablets. Observing reality in the spectral regions of ultraviolet, near infrared and thermal infrared allows students to give a new meaning to the concepts of vision and heat and also to make many aspects, of calorimetry and thermodynamics related to heat transmission, "visible".
- The third step is made of five (4-h each) afternoon (after school) lab lessons at university in which also the physics teacher of the students is present. In the first, students discuss the way and the why a scientific theme has been dramatized the way it is in some physics shows, starting from (but not limited to) "Light from stars" that they have already seen (Fig. 3). A publication created for this purpose by the EU project Teaching Inquiry with Mysteries Incorporated (TEMI) to which the Milan group participated, and concerning the show "Light mystery"—again by "lo Spettacolo della Fisica"—is also used (http://teachingmysteries.eu/wp-content/uploads/2016/04/LIGHT-MYSTERY-Theatre-Script-supporting-science-teaching-v3.pdf).

The second lab lesson is dedicated to providing students with theatrical instruments, starting students with theatrical language, the use of their body and their voice and, in general, with basic elements of a "theatrical grammar".

The third and fourth lab lessons are dedicated to direct experimentation. Students divided into groups are asked to choose one or two themes related to UV and IR vision, to deconstruct the scientific theme and work to bring out various aspects of it in a theatrical way.

In the last lab lesson, students present their original work and discuss it with the other students of the lab and the lab teacher.

3.3 Comments and Perspectives

The attitude of the students in performing dramatization activities proved to be positive in most cases. Many students reported having enjoyed and excited during the activity, and some changed their idea from negative to positive on the scientific subject and appreciated the fact of seeing the same theme from several points of view. Even those who had no affinity with physics developed interest and felt involved. They felt to have learned something with a meaningful learning. But we in Milan believe we have to go even further and are working and planning to strengthen the theatrical activities with the formal ones at school with the creation and experimentation of a tutorial for developing lessons with lab activities about UV and near and far IR, as well as about the vision of colours (with the help of the master students Martina Mulazzi and her tutor in Bologna, Olivia Levrini). In fact, our experience lasting 15 years, with 7 dramas written and performed in more than 400 replicas, strongly suggests that in bridging the gap that is often present among formal, non-formal and informal education, and scientific theatre can be really and extremely useful, but schools need a strong support to make students experience grow also in a formal meaningful way.

4 Active Learning in Teacher Preparation for Active Learning in High Schools (Based on the Presentation of Sergej Faletič at the Symposium)

Research shows that interactive engagement methods lead to better student learning gains than traditional methods (see for example Hake 1998; Freeman et al. 2014; Von Korff et al. 2016; Waldrop 2015). In most cases, especially in high schools with a population of students with very varied interests, the active engagement needs to be provided by the teacher. Therefore, we want to prepare our future teachers to use active engagement methods in their class. We choose investigative science learning environment (ISLE), because it actively engages students in all phases of learning/teaching process, and it is also an epistemologically authentic approach. ISLE engages students in knowledge-generating activities that mimic the actual practices of physics, using the reasoning tools that physicists use when constructing and applying knowledge (Etkina 2015; Brookes et al. 2020).

Through the last decade, we gradually reformed the following courses at our department using ISLE as a theoretical framework: Didactics of physics (sequence of three one-semester courses), communication of physics, methodical practicum, project work in science, How things work and Project laboratory. All courses are either compulsory or elective courses for pre-service physics teachers.

4.1 The ISLE Framework

The ISLE-investigative science learning environment framework has been described in detail elsewhere (Etkina 2015; Brookes et al. 2020; Etkina and Van Heuvelen 2007; Etkina et al. 2020). It places strong emphasis on the epistemology of physics mimicking the process of how scientists acquire new knowledge (see Fig. 4). First, students make an observation of a phenomenon. Next, they design an observational experiment, which will help them observe patterns in the phenomenon. The patterns can be qualitative or quantitative. Then students propose explanations (hypotheses, models) for the observed patterns. The explanations can be either causal (relating

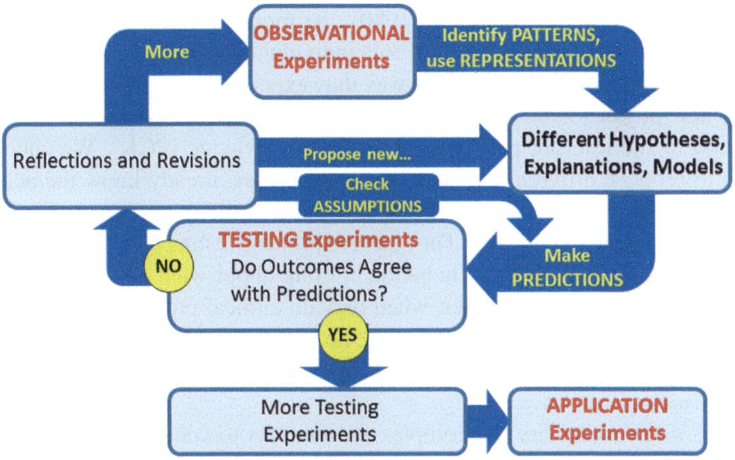

Fig. 4 Simplified diagram of the ISLE process

cause and effect without referring to or describing a specific mechanism) or mechanistic (explaining the mechanism behind the phenomenon). The next phase is to test the explanations by proposing testing experiments, which should aim to reject one or more of the proposed explanations. In case of more explanations, the testing experiments should differentiate between them. It is crucial to predict the outcome of the experiments based on the proposed explanations before performing the experiment. If the outcome of an experiment does not agree with the prediction based on an explanation (as indicated by the "No" in the yellow circle in Fig. 4), then the explanation needs revision: either the assumptions are invalid, and a new prediction can be made based on new assumptions, or the assumptions are valid, and a new explanation has to be proposed, or there is too little data to propose a new explanation and more data has to be gathered by more observational experiments. The process is repeated until only one explanation remains that could not be rejected by the testing experiments. This is accepted as the best explanation so far. Then students can use this explanation (knew knowledge) to apply it in solving practical problems (application experiments). The description is simplified, but it is important to note that the process is neither linear nor cyclic.

4.2 Didactics of Physics

Didactics of physics is a three-level course. Using the ISLE framework, we reformed both lessons and recitations to be as actively engaging as possible. Our goal is to teach the way we would like our students to teach when they go to schools. Over the past 6 years, we have gradually transformed our lessons so that now we are using ISLE approach in every lesson. Before beginning each lesson, we carefully create the "need to know" (Knowles 1980), the motivation for learning the topic. Students work in small groups, solving both individual and group tasks following the ISLE process as described above. This way they experience the process of gradually building new physics knowledge (such as Newton's laws, conservation laws, etc.) in addition to the relevant pedagogical content knowledge (PCK). We encourage students to propose different explanations, even if they already know the accepted one. This helps them practice the step that is most challenging and motivating for students—testing the hypotheses. They have to propose testing experiments to reject some of the explanations, which helps them think about what testing experiments their students might propose. Later, when they do clinical practice in schools, they often realize that high school students come up with similar ideas. We insist that the most important findings come from the class before being reiterated and structured by the instructor.

In cases when topics are too complex for students to come up with the correct explanations, we start from observations and identifying patterns but then create the need to know before "time for telling" (Schwartz and Bransford 1998).

In recitations, students also work in small groups. They solve problems on their own, often using the think-pair-share method, with the instructor providing guidance and assistance. Depending on the type of problem, students start solving the problem alone, before discussing with peers. The final solution is a consensus of the entire group, which distributes the responsibility and builds confidence in the solution. The type of problems reflects the ISLE philosophy as well. Since students are future teachers, many tasks involve evaluating solution or statements given by fictitious persons. These include common ideas and difficulties that students typically have. Students have to recognize productive ideas, even when embedded in incorrect answers and differentiate them from unproductive ideas. This way they develop tolerance and the ability to listen to others' ideas. Students are often asked to suggest an intervention describing how they would respond in a particular situation. We encourage pre-service teachers to think what steps their students should do to come up with a correct understanding rather than them presenting the correct answer or explanation.

We also changed the way we assess students' work. We put large emphasis on formative assessment and allow students to revise and improve their work without being punished for revisions by getting a lower grade (Brookes et al. 2020). Tasks are such that they reflect the emphases of the ISLE method, and students are given a short time (typically less than 24 h) to improve their work after we return them their tests giving them only an overall score without comments on what they need to improve. For the improved solution to be accepted, students need to identify what parts of their work are problematic (they are allowed to consult their peers) and write for each part the following: What did I do incorrectly, How to do it correctly, Why did I make the mistake and How did I learn to do it correctly.

Student practice is an essential component of every teacher preparation course. Our students participate in three types of practice: microteaching (also known as simulation of school lesson), sheltered practice in school (similar to Japanese lesson study (Yoshida 1999)) and clinical practice in school.

In microteaching, student pairs prepare and conduct a micro-lesson using the textbook college physics (Etkina et al. 2019a) and the accompanying book of activities, the active learning guide (Etkina et al. 2019b) that are both following the ISLE approach to all topics. The focus of microteaching is in developing abilities to identify opportunities in the classroom, listen and pay attention to students' ideas and react productively, while using tested material from the textbook and the active learning guide. We emphasize the role of experiments, especially testing ones, in the epistemology of physics. We developed rubrics for microteaching, using the scientific abilities rubrics developed by the Rutgers group[3] (Etkina et al. 2006). Rubrics help students prepare their lessons and allow us to evaluate them.

The sheltered practice has been reformed so that students plan a series of lessons, usually covering an entire topic, where each lesson is designed following the ISLE framework but also taking into account the national curriculum. This way, students have the opportunity to see how a series of lessons come together and gain a better understanding of the entire approach. The students then carry out the entire series of lessons in a real class of high school students, with pairs of students carrying out each lesson. The other students from the programme, the teaching assistant and the course leader observe each lesson from the back without interfering and afterwards we have a thorough analysis of the lesson.

In clinical practice, students spend five weeks in a school and have to carry out at least 13 independently conducted lessons, some of which are observed by the course leader or the teaching assistant. We try to place students with mentoring teachers that are familiar with ISLE and motivated to improve their teaching style.

The Didactics of physics course is complemented with the methodical practicum course, which focuses on the technical and pedagogical aspects of experiments. In this case, using the ISLE framework enables us to go beyond mere manipulation of experiments. Students do projects with an emphasis on familiarizing themselves with contemporary technology; however, they have to propose how each experiment can be used in the classroom and provide a clear description of the learning goals and the role of the experiment in achieving these goals.

5 Pre-service Science Teacher's Experiences of Inquiry-Based Learning in Socio-scientific Contexts (Based on the Presentation of Eilish McLoughlin at the Symposium)

UNESCO's recent report "Rethinking Education: Towards a common global goal?" (UNESCO 2015) reminds us that the changes that we face in the world today are characterized by new levels of complexity and contradiction. Today's citizens need a

[3] Scientific Abilities Homepage, https://sites.google.com/site/scientificabilities/rubrics, last accessed 2021/01/13.

deeper understanding of global societal challenges and their implications for themselves, their families and their communities. This requires a broader vision of an active, engaged and responsible citizenship for the twenty-first century, as described in the objectives of the framework for science education for responsible citizenship (Hazelkorn et al. 2015). These objectives are further highlighted in the OECD Education 2030 framework which aims to build a common understanding of the knowledge, skills, attitudes and values necessary to shape the future towards 2030.[4] In order to equip today's learners with agency and a sense of purpose, and the competencies they need, to shape their own lives and contribute to the lives of others, we need to make changes to science education curricula, pedagogy and assessment practices and support science educators in embedding such approaches in their classroom.

5.1 Theoretical Basis

Strategies for active learning, as described by the GIREP Thematic group highlights that (https://www.girep.org/thematic-groups/strategies-for-active-learning/):

> In order to effectively learn Physics students should do more than just listen to a lesson. They should be engaged in actively reading, writing, posing and discussing questions, gathering data from different sources, building models and in solving problems ultimately aimed at developing their knowledge, skills and attitudes. In Active Learning students are basically engaged in two aspects: doing things and reflecting about the things they are doing.

Inquiry is widely regarded as an effective approach to engaging learners in active learning and developing the learner's scientific knowledge and skills (Bevins and Price 2016). Inquiry teaching is student-centred, and teacher-student collaboration is a central feature of the learning environment. The role of the teacher is to facilitate learning by asking questions and encouraging students to reflect on their current understanding. An inquiry-based learning (IBL) in school science is based around students carrying out investigations including experimentation and secondary research. Inquiry in the context of socio-scientific issues (SSI), which are contemporary and relevant scientific topics with moral or economic implications (Sadler 2009) has further been promoted to enhance student's scientific literacy. SSIs centre on a range of scientific, social or moral viewpoints, which may conflict with the students' own views and thus makes them personally relevant to students (Zeidler and Nichols 2009). This study presents a qualitative analysis of the impact of this active learning approach with physics pre-service teachers (PSTs) to address two research questions:

- What are the PSTs' experiences of carrying out inquiry in the context of SSI as learners?
- What are the PSTs' experiences of carrying out inquiry in the context of SSI as teachers?

[4] OECD Future of Education and Skills 2030 Homepage, http://www.oecd.org/education/2030/learning-framework-2030.htm, last accessed 2021/01/13.

5.2 *Methodology*

This research presents a study of the implementation of a series of workshops with 17 physics pre-service teachers (PSTs). The PSTs were in their second year of a four-year undergraduate BSc in Science Education programme which would qualify them to teach from physics and mathematics or chemistry or teaching at secondary level. The workshops were facilitated at the end of semester two, with the PSTs having completed their first in school teaching placement (total of four weeks) at end of semester one.

The aim of this study was to develop the skills and knowledge of the PSTs as learners and develop their pedagogical approaches as teachers—to order to prepare them to use inquiry-based learning (IBL) approaches in SSI contexts as part of their teaching (Chadwick 2018). The PSTs' co-planned and presented on a series of lessons that incorporated inquiry learning in their own SSI context. The learning outcomes of sessions focussed on PSTS developing the skills of planning for engagement in science including identifying investigable questions, anchoring ideas/ phenomena and preparing casual explanations. PSTs were presented with one of six SSI contexts—gas laws, human vision, satellite communication, modern bridges, rail travel and wind energy.

After the second workshops, PSTs were facilitated to reflect individually on their own learning experiences in terms of the skills of collaboration, communication and time-management. At the end of the workshop series, PSTs were asked to reflect on what was challenging about collaborating with their group to complete and present this plan.

5.3 *Findings*

Analysis of PSTs first reflection revealed that PSTs were engaged in actively involved in the tasks of reading, writing, posing and discussing questions, gathering data from different sources and building models to support their development of content knowledge. However, it was evident that PSTs were easily able to report on what they did but needed additional support to develop their reflective practices. This is essential for future teachers, as Shulman (1986, 1987) states that deliberate reflections are needed in order for teachers to start developing their pedagogical content knowledge (PCK).

At the end of the workshops, the PSTs showed evidence of having developed their own scientific knowledge and skills. In their reflections, they discussed what changes they would make to their classroom practices to extend and deepen student learning, e.g. how they would change the assessments used, how they would scaffold learning to develop conceptual understanding and what experimental investigations they would include. In particular, they expressed that they had increased understanding of the importance of reflection, e.g. one PST stated "*I would add a section to teach the students how to reflect, at the time of writing the plan I didn't think in enough depth about the student's prior reflection skills*".

6 Discussion and Final Remarks

Although the four contributions to this paper deal with different themes, they all give a wide-range view of active learning foundations, methodology and efficacy and are strongly grounded on research on these topics.

Particularly, the first contribution acknowledges that effectiveness of learning ultimately lies in the student's mind being active in producing a response on a short time scale to an intellectual stimulus. This is particularly true of higher education in physics where the aims include the achievement of a deep understanding of cognitively challenging concepts and methods which go far beyond the accumulation of a large amount of factual information and skills; thus, implying that thinking is much more important than remembering or skill in accessing a large factual knowledge base. Research on the mental processes involved in learning and of student different learning styles is promising and can give useful hints to the educators and researchers interested in deepening this subject to improve their understanding of active learning pedagogical and psychological foundations. Also, a better understanding of situations where active learning shows to be not so effective with special students can be obtained, as active learning approaches should be adapted to the specific learning styles and psychological attitudes of those kind of students.

As the literature has shown in the last years, there are many active learning approaches, and research is still needed to well understand their real efficacy in improving student conceptual knowledge and skill development. The second contribution discusses a novel approach to active learning, based on the involvement of students in scientific theatrical activities. Here, the idea is that scientific theatre can be an extremely useful tool to stimulate motivation by arousing positive emotions. In the theatrical approach to physical themes, students first come into contact with the scientific theatre with the direct experience of one or more physics shows, then discuss with physicists and/or theatre professionals the way in which a scientific theme can be dramatized, actively working on a chosen physics topic to bring out various aspects of it with theatrical methods. These activities prove to be effective in producing positive reactions in student attitudes towards science. Many students report being enjoyed and excited during the activities, some appreciated the fact of seeing the same theme from several points of view. Even those who had no affinity with physics developed interest and felt involved. They felt to have learned something with a meaningful learning.

While research on active learning has often focussed on student needs and difficulties, it is undeniable that a focus on the role of the implementer of the active learning approaches, i.e. of the teacher, is at least equally needed. Many research results (Mitchael 2007; van den Berg 1999; Adamson et al. 2003) highlight the difficulties teachers face when they are fronted with the need to take active learning to their classes. A lack of self-confidence with the new pedagogical methodologies typical of active learning and/or lack of comfort with the content to develop, time constraints in development of the new topics/methodologies, doubts about evaluation, personal beliefs on teacher/student role in teaching/learning and on nature of science are only some examples of these difficulties. They may be due to a traditional initial teacher education and must be fronted in order to improve active learning-based teaching, both from a pre-service and an in-service teacher education point of view.

The last two contributions to this paper deal with this aspect. In the third contribution challenges in the implementation of active learning methods in a way that teachers will be able to use in schools are also discussed, mainly in the view of future teacher personal beliefs about what teaching looks like. These beliefs are, in many cases, much different from what is required to effectively implement active learning methods. The contribution describes an implementation of active engagement methods used in high school physics teacher preparation courses, pointing out that for effective implementation, it is crucial to find a well-researched and tested active engagement approach. To this aim, the well-known investigative science learning environment approach is chosen as a theoretical framework. It clearly emphasizes experiments and observations, which are well acknowledged as crucial for the epistemology of physics. It is also shown that looking at the experiments as observational, testing and application experiments helps students to think like scientists and teachers to make progress towards a real understanding of active learning meaning.

The focus of the research discussed in the fourth contribution is on pre-service science teacher's experiences of inquiry-based learning. Particularly, a study reporting on an inquiry-based learning (IBL) approach carried out in context of socio-scientific issues (SSI) with a cohort of pre-service physics teachers (PSTs) is discussed. The approach aims at developing the pre-service teacher's knowledge and skills for adopting inquiry and SSI contexts in their classroom practices. The PSTs are facilitated to carry out their own inquiry and reflect on their experiences of the pedagogical approach and context of the SSI. The PSTs' experience as teachers showed an increased awareness of the advantages of more student-centred active learning approaches and the need for teachers to have well-developed PCK with integrated knowledge of theory, practice and reflection (Juhler 2018). This study highlights that deliberate reflections are needed by learners so they are not only active in doing things, but also active in reflecting about the things they are doing.

In conclusion, the four contributions here presented show that, despite the consistent amount of literature results related to active learning, research on this topic is still scientifically interesting for both educational researches and teachers. Closer attention should be deserved to aspects like the need for a better understanding of the pedagogical and psychological foundations of active learning, for deepening possible new ways to do "active learning", possibly involving socially well-recognized channels, like theatre, and for more focus on teacher pre-service and in-service education aimed at helping teachers to effectively shape their beliefs on the teaching/learning processes and to properly appropriate of methods and contents fostering active learning.

References

Adamson SL, Banks D, Burtch M, Cox III R, Judson E, Turley JB, Benford R, Lawson AE (2003) Reformed instruction and its subsequent impact on secondary school teaching practices and student achievement. J Res Sci Teach 40(10):939–957

Akerib DS et al (2012) LUXSim: a component-centric approach to low-background simulations. Nucl Instrum Methods Phys Res Sect A 675:63–77

Andersen T (1993) See and hear, and be seen and heard. In: Friedman S (ed) The new language of change: constructive collaboration in psychotherapy. Guilford Press, New York, pp 303–322

Barbieri S, Carpineti M, Giliberti M, Rigon E, Stellato M, Tamborini M (2015) Good vibrations: a workshop on oscillations and normal modes. Il Nuovo Cimento 38:86

Bevins S, Price G (2016) Reconceptualising inquiry in science education. Int J Sci Educ 38(1):17–29

Bonwell CC, Eison JA (1991) Active learning: creating excitement in the classroom. ASHEERIC higher education report no. 1. George Washington University, Washington, DC

Brookes D, Etkina E, Planinsic G (2020) ISLE: implementing an epistemologically authentic approach to student-centered inquiry learning. Phys Rev PER 16:020148

Carpineti M, Cavallini G, Giliberti M, Ludwig N, Mazza C, Perini L (2006) Let's throw light on matter: a physics show for primary school. Nuovo Cimento B 121(8):901–911

Carpineti M, Giliberti M, Ludwig N (2016) Theatre as a means of supporting the teaching of science—light mystery—a script with added comments. ISBN/EAN: 978-94-91760-19-8

Chadwick R (2018) Development and assessment of scientific literacy for secondary level science education. PhD thesis, Dublin City University

Chi MTH, Adams J, Bogusch E, Bruchok C, Kang S, Lancaster M et al (2018) Translating the ICAP theory of cognitive engagement into practice. Cogn Sci 42:1777–1832

EPS Physics Education Division (2012) Position paper on physics education. Europhys News 43(5):05–07

Etkina E (2015) Millikan award lecture: students of physics—listeners, observers, or collaborative participants in physics scientific practices? Am J Phys 83:669–679

Etkina E, Van Heuvelen A (2007) Investigative science learning environment—a science process approach to learning physics. In: Redish EF, Cooney P (eds) Research based reform of university physics, (AAPT). Online at http://per-central.org/per_reviews/media/volume1/ISLE-2007.pdf

Etkina E, Van Heuvelen A, White-Brahmia S, Brookes DT, Gentile M, Murthy S, Rosengrant D, Warren A (2006) Scientific abilities and their assessment. Phys Rev ST Phys Educ Res 2:020103

Etkina E, Planinsic G, Van Heuvelen A (2019a) College Physics. Explore and apply, 2nd edn. Pearson Education, USA

Etkina E, Brookes D, Planinsic G, Van Heuvelen A (2019b) College Physics active learning guide, 2nd edn. Pearson Education, USA

Etkina E, Brookes DT, Planinsic G (2020) Investigative science learning environment: learn physics by practicing science. In: Mintzes JJ, Walter EM (eds) Active learning in college science: the case for evidence based practice. Springer, Cham, pp 359–383

Fazio C (2020) Active learning methods and strategies to improve student conceptual understanding: some considerations from physics education research. In: Guisasola J, Zuza K (eds) Research and innovation in physics education: two sides of the same coin. Springer, Cham, pp 15–35

Freeman S, Eddy SL, McDonough M, Smith MK, Okoroafor N, Jordt H, Wenderoth MP (2014) Active learning increases student performance in science, engineering, and mathematics. Proc Natl Acad Sci 111(23):8410–8415

Hake RR (1998) Interactive-engagement versus traditional methods: a six-thousand-student survey of mechanics test data for introductory physics courses. Am J Phys 66:64–74

Hazelkorn E, Ryan C, Beernaert Y, Constantinou CP, Deca L, Grangeat M, Karikorpi M, Lazoudis A, Casulleras RP, Welzel-Breuer M (2015) Science education for responsible citizenship. Report to the European Commission of the expert group on science education

Juhler MJ (2018) Pre-service teachers' reflections on teaching a physics lesson: how does lesson study and content representation affect pre-service teachers' potential to start developing PCK during reflections on a physics lesson? NorDiNa 14:22–36
Knowles MS (1980) The modern practice of adult education. Cambridge Adult Education, Englewood Cliffs, NJ
Mitchael J (2007) Faculty perceptions about barriers to active learning. Coll Teach 55(2):42–47
Redish EF (2014) Oersted lecture 2013: how should we think about how our students think? Am J Phys 82:537–551
Sadler TD (2009) Situated learning in science education: socio-scientific issues as contexts for practice. Stud Sci Educ 45:1–42
Schwartz DL, Bransford JD (1998) A time for telling. Cogn Instr 16:475–522
Shulman L (1986) Those who understand: knowledge growth in teaching. Educ Res 15(1):4–14
Shulman L (1987) Knowledge and teaching: foundations of the new reform. Harv Educ Rev 57(1):1–22
Turing AM (1950) Computing machinery and intelligence. Mind 49:433–464
UNESCO (2015) Rethinking education: towards a common global goal? http://unesdoc.unesco.org/images/0023/002325/232555e.pdf
van den Berg R (1999) The permanent importance of the subjective reality of teachers during educational innovation: a concerns-based approach. Am Educ Res J 36(4):879–906
Von Korff JB, Gomez KA, Heckendorf T, McKagan SB, Sayre EC et al (2016) Secondary analysis of teaching methods in introductory physics: a 50 k-student study. Am J Phys 84(12):969–974
Waldrop MM (2015) The science of teaching science. Nature 523:272–274
Yoshida M (1999) Lesson study: a case study of a Japanese approach to improving instruction through school-based teacher development. Doctoral Dissertation, University of Chicago
Zeidler DL, Nichols BH (2009) Socioscientific issues: theory and practice. J Elem Sci Educ 21:49–58

Innovative Projects

Max's Worlds: An Innovative Project for K-6 Science Education

Federico Corni and Liliana Dozza

Abstract Max's Worlds is a laboratory project aimed at developing a coherent vertical curriculum suitable for kindergarten and primary school teachers, either for their pre-service university program or for in-service training activities. The project provides didactic suitcases: after introducing the project theoretical foundations, the paper will focus on the energy suitcase, its content, and reaction of student teachers after laboratory activities using the provided materials.

Keywords Energy · Embodied mind · Vertical curriculum

1 Introduction

Max's Worlds is a science project ideated and developed in the framework of MultiLab at the faculty of education of the Free University of Bolzano, Italy. It aims to develop a coherent, vertical curriculum for primary physics education, in continuity with kindergarten and middle school.

The coherence of the curriculum is given by the philosophical and disciplinary foundation characterizing the whole project. The verticality of the curriculum is ensured by the methodological choices of the project. The continuity of the curriculum is given by the organization of the didactic materials provided by the project. They are collected in thematic suitcases, available for schools of every grade. A suitcase contains the materials for pupil activities, the teacher guide, and the digital products accompanying the didactic path. The didactic paths supported by the suitcases are narrative and strictly analogical, building on the same set of elementary concepts. At present, four Max's World suitcases are available (Fig. 1): HydroLand (fluid phenomena), ThermoLand (thermal phenomena), ElectroLand (electric phenomena), and ErgoLand (energy).

F. Corni (✉) · L. Dozza
Faculty of Education, Free University of Bozen-Bolzano, Via Ratisbona 16, 39042 Bressanone, Italy
e-mail: federico.corni@unibz.it

B. Jarosievitz and C. Sükösd (eds.), *Teaching-Learning Contemporary Physics*, Challenges in Physics Education,
https://doi.org/10.1007/978-3-030-78720-2_16

Fig. 1 Max's Worlds suitcases: (from the left) ElectroLand, ErgoLand, ThermoLand, and HydroLand

Due to the integration of science and humanities, Max's Worlds has a strong interdisciplinary character, which allows teachers to easily introduce it into their didactic planning. Thus, science is linked to the other school disciplines with a fruitful and effective mutual enrichment.

In this paper, we will introduce the theoretical foundations of Max's Worlds (Sect. 2). As an example, Sect. 3 will describe the content and activities of ErgoLand,[1] the energy suitcase. We want to answer the question whether this narrative and interdisciplinary approach is effective for student teachers in terms of learning physics as well as awareness and attitude toward physics. To do so, Sect. 4 will show the results derived from experimentation of laboratory activities in the master degree in primary education at the Free University of Bozen-Bolzano in Bressanone and at the University of Modena and Reggio Emilia in Reggio Emilia (Italy).

2 Theoretical Foundations of Max's Worlds

The project relies on the embodied mind theory, according to which the sources of conceptualization are derived from the abstractions developed by the mind on the basis of sensory-motor experience (Johnson 1987; Lakoff and Johnson 1999). Image schemas (Johnson 1987; Croft and Cruse 2004; Evans and Green 2006; Hampe 2005) are the most elementary abstractions that we project onto experience to understand

[1] Ergoland suitcase is at the base of the coordination and support action H2020 European Project FCHgo (Fuel Cell HydroGen educatiOnal model for schools, Call H2020-JTI-FCH-2018-1 n. 826246), fchgo.eu, under the Fuel Cell and Hydrogen Joint Undertaking (FCH JU). Purpose of FCHgo is to encourage—for the coming generations—a culture of ecological awareness and to develop behaviours based upon sound knowledge of key technologies. It aims at creating an educational program delivery model (EPDM) with educational material for young learners (primary and secondary schools) and their teachers to be used in primary and secondary schools about science and technology of hydrogen fuel cells.

the world. Examples of image schemas are VERTICALITY, CONTAINER, PATH, and SUBSTANCE. Linguistically speaking, this is to say that metaphors and metaphoric expressions are largely, and often unconsciously, present in our discourses and explanations (Lakoff and Johnson 1980). A metaphor is a projection of a known concept onto an unknown or complex concept we want to (partially) understand. In this view, scientific language is metaphorical too (Amin 2009; Fuchs 2010; Lancor 2014). We say for example that temperature is falling (referring to the metaphor TEMPERATURE IS A VERTICAL SCALE) that a body loses heat (referring to the metaphor BODIES ARE CONTAINERS OF HEAT, and HEAT IS A FLUID(-LIKE) SUBSTANCE), or that electricity flows through the wires (referring to the metaphor ELECTRICITY IS A FLUID(-LIKE) SUBSTANCE and A WIRE IS A PATH). The project focuses on a small set of elementary concepts (image schemas) used as the source of employed (figurative) language, allowing subjects to be treated in natural language, characterized by the frequent recourse to analogies. Main concepts are FLUID (-LIKE) SUBSTANCE, VERTICALITY and VERTICAL SCALE, FORCE/POWER, CONTAINER, and PATH.

Continuum physics and modern macroscopic physics best use such fundamental and figurative concepts coherently (Fuchs 2010). Working on these principles, Fuchs (2015) introduced the notion of force of nature (FoN). Every FoN (fluids, electricity, motion, heat, and gravity) has its own extensive quantity (FLUID-LIKE SUBSTANCE: volume, electric charge, momentum and angular momentum, entropy, and gravitational mass), with a law of balance (evidencing whether it is conserved or produced and/or destroyed). The conjugated intensive quantities or potentials (VERTICAL SCALE: pressure, electric potential, velocity and angular velocity, temperature, and gravitational potential) are the driving forces (difference of potential) for flow (current). Constitutive relations link these quantities and introduce the notion of capacitance (CONTAINER) and resistance (PATH). Complex phenomena always involve more than one force of nature: one force changes (decreasing its potential) and causes another force to change (increasing its potential). Energy (FORCE/POWER) is the quantity that accounts for these interactions and that regulates both the quantities and the potential changes involved.[2]

The FoN treated in the didactic suitcases (i.e., water, electricity, heat), as well as energy, are offered to pupils of all grades relying on analogical reasoning. Thus, for example, a container is filled with water, a capacitor is charged, and a metal cube is heated, while pressure, electric potential, and temperature are monitored, respectively. In the case of circuits, water current, generated by a pump, drives a small fan; electric current, given by a generator, makes a light bulb shine; and heat current through a metallic rod, between hot and cold surfaces, drives a Peltier device (Fig. 2). Parallel and series circuits are possible.

The age differentiation is on the plane of the cognitive tools attributed to children, according to Egan's theory of the understandings (Egan 1988, 1990, 1997, 2005). Among the others, the main cognitive tool the project employs is the story (Fuchs

[2] On the same theoretical foundations, a narrative approach to mechanics for secondary school students has been developed and is in use in the educational laboratory of the Ducati Foundation in Bologna (Corni et al. 2018).

Fig. 2 Analogical experiments proposed in the Max's Worlds suitcases. Top: a CONTAINER is filled with FLUID-LIKE SUBSTANCE and increases its intensity (VERTICALITY). Bottom: in a circuit, current, maintained by a pump, drives a device (FORCE/POWER)

2013, 2015; Corni 2014; Fuchs et al. 2020). The metaphorical expressions in the natural language needed to describe a phenomenon or a chain of phenomena, interlaced narratively, can lead to stories of FoN (Fuchs 2015), where the characters are neither humans nor humanized entities, but the FoN themselves (water, etc.) that act and suffer in natural processes (see for example, the Perpetuum Mobile story below). Stories can be seen as cognitive tools for our imagination underlying our knowledge of the world (Egan 1997).

For a more extensive and exhaustive treatment of the foundations of the project, see Corni and Fuchs (2020).

3 ErgoLand: Max's Worlds Energy Suitcase

To show how a topic is treated in the Max's Worlds project, the content and some activities related to the energy suitcase ErgoLand will be described. We are interested in answering the question to what extent this narrative and interdisciplinary approach is effective for student teachers in terms of learning physics, i.e., the energy concept, as well as of awareness and attitude toward physics.

3.1 Content of ErgoLand

The materials contained in the suitcase are conceived either for teachers (in pre-service and in-service training), or for pupils, according to their school level (K: Kindergarten; P: Primary School; M: Middle school):

1. Song: «Discover the world with Stella» (K, P)
2. Big book: «Stella goes to kindergarten» (K)
3. Story movie: «Stella to the discovery of the world» (P)
4. Story movie: «Perpetuum Mobile» (P, M)
5. Assembly boxes (commercial): Jet car, dynamo torch, wind-mill generator, solar car, water rocket, hydrogen car, … (P, M)
6. Play cards (P, M)
7. Process diagram cards (M)
8. Process diagram software (M)
9. Tablet app of augmented reality (M).

Only the materials suited for primary school level (P) will be described here.

«Discover the world with Stella» (1.) is a melodic and rhythmic song, sang by Stella, the character who accompanies the children in every activity with suggestions and questions.

The story «Stella to the discovery of the world» (3.) is a movie divided into three acts that leads the pupils' through the discovery and application of the concept of energy in various simple toys (jet car, dynamo torch, wind-mill generator, solar car, water rocket, and hydrogen car) offered in assembly boxes (5.).

The «Perpetuum Mobile» movie[3] (4.) is at the core of the ErgoLand materials, designed to introduce the visual and verbal metaphors for energy (Corni et al. 2019a). It presents an allegory of the Earth as an open flow system. An inventor attempts to create a perpetuum mobile consisting of a series of devices in a circle: a lamp, a solar cell, an electric water pump, a water wheel, and a generator feeding back to the lamp. The working quantities between these devices (electricity, light, water, and motion) are symbolized as agent and patient spirits that carry (agents) and hand over energy—represented as dust—as the processes in the machine are running. In every exchange, some energy is lost and causes the production of heat (symbolized as a snake spirit). At the end, to make the machine work, energy losses are replaced by an external source, the Sun. The graphics of the movie delivers several visual metaphors, mainly the FLUID-LIKE SUBSTANCE for energy and working quantities, HIGH and LOW LEVEL for the potentials.

Finally, the decks of 90 cards (6.) display 15 different energy carriers at different levels of intensity on a scale 1–3. Instructions are supplied for simple games that convey the energy conceptual aspects in their rules.

---

[3] M. Deichmann's thesis. Im ubertragenen Sinne. Metaphern und Bildvergleiche in der Wissern-schaft. Zurcher Hochschule der Kunste, Zurich; 2014. Available from: http://vimeo.com/983 11515.

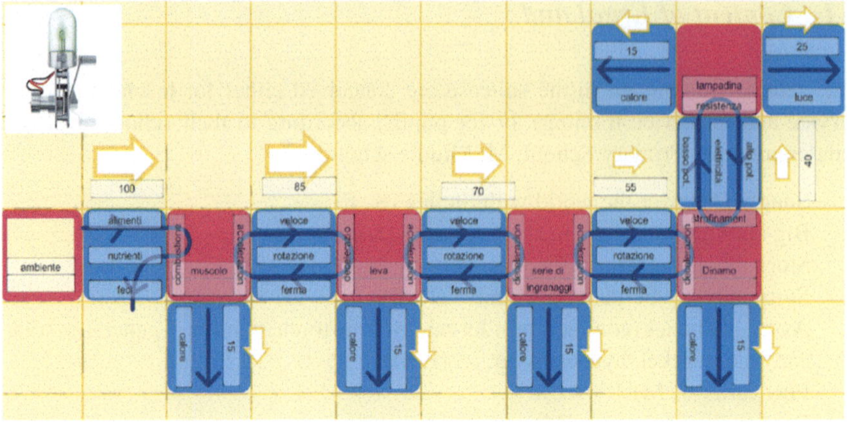

Fig. 3 Process diagram of the functioning of the dynamo torch, as it can be built with the ErgoLand's software. FoN (blues arrows) interact through the coupler devices (pink boxes) and make energy flow (yellow big arrows)

In addition, process diagrams are briefly tackled. They can be created using the provided cards and software (7., 8.), and they are suitable for middle school pupils. They have been used with in the experimentation with student teachers we are going to present in this paper. For further details, please refer to previously published papers about process diagrams (Fuchs 2010; Corni et al. 2019b). A process diagram is a schematic way of rendering a story of interaction within FoN in a technical (or even in a natural) device. Figure 3 shows an example of process diagram related to the functioning of the Dinamo torch toy, developed by a group of student teachers. Here, chemical substances or food, angular momentum, electricity, light and heat interact. In a coupler allowing the interaction (pink box), the potential of one FoN drops, and energy is made available to the next FoN, whose potential rises. Part of energy is also lost due to heat production. The labels in the blue boxes allow students to name the FoN (central label), its high or low potential state (top and bottom labels), and the processes of potential increase and decrease (left and right vertical labels). For example, angular momentum can be at high or low angular speed, due to angular acceleration (or torque) or deceleration (or friction).

3.2 Activities with ErgoLand

The first activity, suggested by the first act of "Stella to the discovery of the world" movie, consists in playing with a jet car to make it move farther. Here, working in groups, students recognize that the quantity of air in the balloon (amount of extensive quantity) and its pressure compared to the external one (difference of potential) define the energy of the car. The second activity, suggested by the movie, consists in working with toys in assembly boxes. The students have to analyze the "anatomy" (which parts is composed of) and the "physiology" (how the parts work) of the toy. The last act of "Stella to the discovery of the world," that comprises the vision of the "Perpetuum mobile" movie, requires the students to find the FoNs that carry energy in the functioning of the toy. Then, they have to create a story about the operation of the toy, telling the interactions of the FoNs. They also have to build a process diagram, and, finally, to project a possible embodied play of the story with pupils. In the play, groups of pupils interpret the forces of nature. Every pupil who interprets a FoN at high potential carries confetti, i.e., energy, and hands them over when he or she meets another pupil in a suitable place representing a coupler. The behavior of pupils should mime the nature of the specific energy carrier. For example, pupils representing electricity can walk in a circle, like electricity in a circuit, or pupils representing light can stand up (light production), move (light propagation), and get sleeping (light absorption).

4 Data Sources, Results, and Discussion of Experimentation of ErgoLand Materials

In the last academic years (since 2017–2018), ErgoLand was used with student teachers in the physics lab provided in the master's program in primary teacher education at the Faculty of Education of the Free University of Bozen-Bolzano (UNIBZ) (25 h) and at the Department of Education of the University of Modena and Reggio Emilia (UNIMORE) (16 h). The lab is part of the integrated course of physics (9 university credits in total) comprising also 60 h of lectures.

To what extent the students have learnt the concept of energy in a significant way for their future profession will be deduced from the documents produced by the students in the academic year 2018–2019 during the lab activities (process diagrams, stories, scripts of plays) (11 groups of UNIBZ students—60 students) and from an anonymous questionnaire given after the exam in the academic year 2018–19 (114 UNIMORE + 39 UNIBZ students = 153 students). Some items of the questionnaire will be used to obtain information about the students' awareness of and attitude toward physics.

4.1 Students' Learning

The process diagrams, stories, and play scripts produced by the groups of students about the functioning of a toy have been analyzed according to the following criteria, where applicable: correctness, number and order of the employed FoNs or energy carriers, nature of the energy carriers (conserved or not conserved), explication of heat production, explication of the energy carrier potential (high or low), explication

Table 1 Percentage of correct answers in process diagrams referred to the expected number of energy carriers present in the toy

Group #	Numb. of energy carrier (%)	Order of energy carriers (%)	Nature of energy carriers (%)	Heat production (%)	High/low potential (%)	Process of incr./decr. potential (%)	Energy conservation
1	100	100	100	100	100	100	Yes
2	100	100	100	100	100	100	Yes
3	100	100	100	100	100	100	Yes
4	100	100	100	100	100	100	Yes
5	100	100	100	100	100	100	Yes
6	100	100	100	100	100	0	Yes
7	100	100	100	100	100	67	Yes
8	100	100	100	100	100	33	Yes
9	100	100	100	100	100	100	Yes
10	100	100	100	100	100	67	No
11	100	100	100	100	100	50	No
Ave	100	100	100	100	100	74	82%

of the process of increasing or decreasing of the potential, and explication of energy conservation.

The percentage of correct answers of the various groups are reported, referred to the expected number of energy carriers in the respective process diagrams: Table 1 shows the results for the students' process diagrams, and Table 2 shows the results for stories and play scripts.

Students showed ability in building process diagrams (Table 1). In particular, they reported 100% of the energy carriers, 100% in the correct order, meaning they were able to find all the expected energy carrier and to differentiate them from couplers (e.g., electricity or light are differentiated from lamps or photovoltaic panels, respectively). Moreover, 100% of energy carriers were correctly recognized as conserved or not conserved, as well as named in their different potential states. Heat production was correctly reported in 100% of cases, too. Some difficulties were encountered in naming the processes of increasing or decreasing potentials, probably due to the required specific language. Two groups did not show the energy arrows, so energy conservation was not explicitly represented.

Students' stories and play scripts deserve a brief discussion. If we compare Tables 1 and 2, it is clear that the percentages corresponding to the same criteria are lower in the former than in the latter. This can be derived by considering that in a story or in a play the narrative language does not allow the same precision and significance as a formalized language, like in process diagrams. Nevertheless, the quality of the stories is good, and in many cases, the addition of a greater specification would have led to a weighting and a distortion of the narrative language. It is worth

Table 2 Percentage of correct answers in stories and play scripts referred to the expected number of energy carriers present in the toy

Group #	Stories				Play scripts			
	Nature of energy carriers (%)	Process of increasing the potential (%)	Process of decreasing the potential (%)	Heat production (%)	Nature of energy carriers (%)	Process of increasing the potential (%)	Process of decreasing the potential (%)	Heat production (%)
1	75	100	75	0	–	–	–	–
2	100	100	50	100	67	100	50	100
3	100	100	100	100	100	100	100	100
4	75	100	75	25	75	100	75	75
5	50	100	50	100	100	100	50	100
6	67	50	33	100	100	100	100	100
7	25	33	25	0	50	67	100	0
8	75	100	50	75	100	100	50	100
9	67	100	100	100	100	0	0	100
10	67	50	100	100	67	50	100	67
11	67	100	50	50	67	100	50	100
Ave	70	85	64	68	83	82	68	84

considering that in the play scripts, the percentages are higher, especially for nature of energy carriers and for heat production, because in a play, the gestural language allows information to be added to the verbal narrative language.

A questionnaire item concerning the disciplinary learning was also filled in. The question was: What are the main characteristics of energy? Every student listed several characteristics (3.7 on average, with standard deviation 1.0) that have been grouped into categories. They are indicated in Table 3. The main characteristics of energy (transfer, storage, conservation, transport, and loss by heat) are present in high percentages, and at least, two of them were reported by all students.

4.2 *Students' Awareness of and Attitude Toward Physics*

The questionnaire given to students included the following questions:

1. What is the basic concept you learned in this course?
2. After this course, has your idea of physics changed compared to the one you had after you graduated? In what sense?
3. In your daily life, have you ever relied on a concept that you learned during the course?
4. Do you think you are able to apply what you have learned in this course in your teaching practice?

Table 3 Energy characteristics categories

It transfers	87%
It accumulates	76%
It is conserved	75%
It is carried	65%
It is subject to a law of balance	64%
It is lost with heat	60%
It is measured in joules	5%
It is an agent	4%
It is a fluid-like quantity	4%
It is released when a potential falls	4%
It can be added	4%
There are different forms of energy	4%
There are not different forms of energy	4%

The answers to question 1 have been divided into two categories depending on whether the students referred to a method of teaching/learning or to a disciplinary topic (Table 4). Their answers refer to the specific features of the Max's World approach, i.e., the central role of FoN and energy, and the use of metaphors, analogies, and narrative language.

The majority of students declared that their image of physics has changed after the course activities (Table 5). Most of them, in secondary school, had difficulties with physics, due to its abstractness and/or to its over mathematization. Very significant

Table 4 Percentage of students' answers to the question: what is the basic concept you learned in this course?

Methodological category		Disciplinary category	
Metaphor and analogy	35%	Energy	40%
Generic	25%	Aspects of forces of nature	26%
Narratives and stories	15%	Heat, entropy, and temperature	12%
Induction from experience	15%	Forces of nature flow	11%
Image schema and language	10%	Forces of nature (generic)	9%
		Conservation	3%

Table 5 Percentage of students' answers to the question: after this course, has your idea of physics changed compared to the one you had after you graduated?

Definitely yes	58%
In some important respects	38%
In some secondary respects	4%
It has not changed	0

Table 6 Categories of students' perception of change in their idea of physics

Reference to an understanding	44%
Physics is concrete	35%
Generic reference to a previous negative view	22%
Physics is simple	15%
Use of experiments	9%
Use of metaphors	7%
Scientific method	5%

Table 7 Percentage of students' answers to the question: in your daily life, have you ever relied on a concept that you learned during the course?

Often	47%
Sometimes	47%
Once	5%
Never	0

Table 8 Percentage of students' answers to the question: do you think you are able to apply what you have learned in this course in your teaching practice?

Definitely yes	20%
Yes	40%
I think yes	11%
I want to try	7%
Probably yes	2%
I hope yes	15%
Partly yes	4%
No	2%

is the reference to understanding (44%) and concreteness (35%) in the perception of the discipline (Table 6).

Tables 7 and 8 refer to the students' perception of relevance of physics to their life and to their future profession. Most students (94%) declared they had several not-secondary-school occasions where they remembered a physics concept they learned. Moreover, they are positive (only 2% are definitely negative) in being able to employ what they learned in their school practice.

5 Summary

This paper presented the project Max's Worlds, based on a narrative approach to physics education relying on metaphorical and analogical reasoning, as well as on stories of forces of nature. In particular, the paper introduced ErgoLand materials, the laboratory experimentation with teacher students in two Italian university sites, and the students' outcomes, in terms of disciplinary learning regarding energy and attitude toward physics.

The students showed their knowledge of the concept of energy in its various features. They also proved to be able to apply its main features in the interpretation of the functioning of toys, similar to those they could use in school. For the course, which includes also the Lab, they had positive reactions, especially their attitude towards physics. Not only did students claim to understand physics, but they also perceived it as useful in their everyday life and for their future professional career.

Acknowledgements This research activity was developed within the Free University of Bozen-Bolzano project n. BW2809 "PPSE: Primary Physical Science Education. Courses and materials for teacher education based upon an imaginative (metaphoric and narrative) approach to the experience of Forces of Nature."

References

Amin TG (2009) Conceptual metaphor meets conceptual change. Hum Dev 52(3):165–197

Corni F (2014) Stories in physics education. In: Sidharth BG, Michelini M, Santi L (eds) Frontiers of fundamental physics and physics education research. Springer proceedings in physics, vol 145. Springer, Cham, pp 385–396. https://doi.org/10.1007/978-3-319-00297-2_38

Corni F, Fuchs HU (2020) Primary physical science for student teachers at kindergarten and primary school levels. Part I—foundations of an imaginative approach to physical science. Interchange 51:315–343. https://doi.org/10.1007/s10780-019-09382-0

Corni F, Fuchs HU, Savino G (2018) An Industrial Educational Laboratory at Ducati Foundation: narrative approaches to mechanics based upon continuum physics. Int J Sci Educ 40(3):243–267

Corni F, Fuchs HU, Landini A, Giliberti E (2019a) Visual and gestural metaphors for introducing energy to student teachers of primary school and kindergarten levels. J Phys Conf Ser 1287:012043. https://doi.org/10.1088/1742-6596/1287/1/012043

Corni F, Fuchs HU, Dumont E (2019b) Conceptual metaphor in physics education: roots of analogy, visual metaphors, and a primary physics course for student teachers. J Phys Conf Ser 1286:012059. https://doi.org/10.1088/1742-6596/1286/1/012059

Croft W, Cruse DA (2004) Cognitive linguistics. Cambridge University Press, Cambridge

Egan K (1988) Primary understanding. Routledge, New York

Egan K (1990) Romantic understanding. The development of rationality and imagination. Routledge, New York

Egan K (1997) The educated mind. The University of Chicago Press, Chicago

Egan K (2005) An imaginative approach to teaching. Wiley, New York

Evans V, Green M (2006) Cognitive linguistics. An introduction. Edinburgh University Press, Edinburg

Fuchs HU (2010) The dynamics of heat, 2nd edn. Springer, New York

Fuchs HU (2013) Costruire e utilizzare storie sulle forze della natura per la comprensione primaria della scienza. In: Corni F (eds) Le scienze nella prima educazione. Un approccio narrativo a un curricolo interdisciplinare. Erickson, Trento, pp 59–84. English version: Meaning in nature—from schematic to narrative structures of science. Retrieved from www.hansfuchs.org/LITERATURE/Literature.html

Fuchs HU (2015) From stories to scientific models and back: narrative framing in modern macroscopic physics. Int J Sci Educ 37(5–6):934–957

Fuchs HU, Dumont E, Corni F (2020) From metaphor to narrative in macroscopic physical science. In: Corni F, Altiero T, Landini A (eds) Innovazione della didattica delle scienze nella scuola primaria e dell'infanzia: al crocevia fra discipline scientifiche e umanistiche. CLEUP, Padova

Hampe B (2005) From perception to meaning. Image schemas in cognitive linguistics. Walter de Gruyter GmbH & Co, Berlin

Johnson M (1987) The body in the mind. University of Chicago Press, Chicago

Lakoff G, Johnson M (1980) Metaphors we live by. University of Chicago Press, Chicago

Lakoff G, Johnson M (1999) Philosophy in the flesh. Basic Books, New York

Lancor R (2014) Using metaphor theory to examine conceptions of energy in biology, chemistry, and physics. Sci Educ 23(6):1245–1267

Creativity Development Through Problem-Based Informal Science: The Case of Double Exceptionality Students

Martha Lucia Orozco Gómez, Radu Bogdan Toma, and Maria Merino

Abstract The development of students' creativity is a critical objective of reform-based science curricula. This study describes an after-school program, called *TECNOArtea*, designed for the development of double exceptionality students' creativity through problem-based learning (PBL) science sessions. Subsequently, the results of a 10-week pilot study are presented. Overall, the findings are positive, revealing that when compared to a wait-list control group, treatment students' creativity significantly improved following participation in the program. This outcome supports the effectiveness of problem-based informal science programs for the development of creativity in students diagnosed with double exceptionality.

Keywords Autism spectrum disorder · Creativity · Problem-based learning

1 Introduction

There is currently a steady increase in the diagnosis of people with autism spectrum disorder (ASD). The educational system must address and provide an effective educational response, so it is necessary to develop an understanding of the cognitive profile of these students and thus ensure successful experiences that promote the insertion into the adult life of this group.

On the other hand, the science curriculum and its delivery to students are undergoing continuous reform worldwide. Nowadays, student-centered teaching approaches are being promoted over traditional pedagogies at all education levels to engage students in scientific and engineering practices (NGSS Lead States 2013).

M. L. Orozco Gómez
Departamento de Ciencias de la Educación, Universidad de Burgos, Burgos, Spain

R. B. Toma (✉)
Departamento de Didácticas Específicas, Universidad de Burgos, Burgos, Spain
e-mail: rbtoma@ubu.es

M. Merino
Asociación Autismo Burgos, Burgos, Spain

B. Jarosievitz and C. Sükösd (eds.), *Teaching-Learning Contemporary Physics*, Challenges in Physics Education,
https://doi.org/10.1007/978-3-030-78720-2_17

There is a need to advance science education beyond the mere learning of scientific concepts to the development of those competencies that allow individuals to become active citizens with the capability to critically engage in problem-solving processes that address societal challenges (European Commission 2015). Among many competencies and skills, creativity is being endorsed as critical for everyday life (Runco 2014), and problem-based learning (Schmidt 1983) is being promoted as an effective active teaching method.

The literature on enhancing and developing creativity through enriched educational settings is scarce (Fekula 2011). This lack is even more pronounced in the population with ASD (Takeuchi et al. 2014). Also, there is an ongoing debate about whether creativity is an innate characteristic or a construct that can be stimulated and fostered without the need for a genetic and neurological basis. Therefore, it is necessary to provide empirical evidence to help fill the existing gap in the literature of evaluation studies that specifically examine the effectiveness of programs aimed at developing ASD student's creativity.

Consequently, the main aim of this study is to present the UBUingenio initiative as an out-of-school science program focused on fostering science learning and development of creativity. Under the UBUingenio initiative, two subprograms destined for gifted students (called *Ingenio*) and for students diagnosed with dual exceptionality (called *TECNOartea*) were designed and implemented at the University of Burgos (Spain). In this paper, we specifically examine the results of a pilot study performed during the 2017–2018 academic year for the *TECNOArtea* program. Within the framework of the *TECNOArtea* program, the aim has been to deepen the understanding of effective pedagogies for the development of creativity as an underlying factor in critical thinking and divergent thinking. This is a particularly valuable aspect for the development of skills aimed at the improving functional academic competences in students with autism. Consequently, the main research question guiding this study was: Is there a significant change in dual exceptionality student's creativity following participation in a problem-based, out-of-school science program?

1.1 Defining Creativity

Creativity is a multifaced and complex construct. Novelty is generally considered to be the main expression of the creative process. In different contexts, this novelty may represent the realization of discoveries, design, and generation of new products, problem-solving, or the development of critical thinking for the resolution of everyday situations of technical, social, or cultural origin (Snyder and Snyder 2008; Piawa 2010; Chan 2013; Marrapodi 2003; Sternberg and Lubart 1995). Despite the many definitions existing in the literature, there seems to be a growing consensus that creativity is related to inventing, designing, or proposing novel work. Therefore, creative people are conceived as those who can produce something new. In this study, we adopt Sternberg and Lubart's (1995) definition of creativity as the ability to generate new, appropriate, and high-quality ideas.

1.2 Autism Spectrum Disorder (ASD)

ASDs have a neurobiological basis, linked to deficits in inter-subjectivity, social attention, executive functioning, mental abilities, and weak central coherence, which affect the person's functionality in the development of communication and social relationships. This impacts on socio-emotional reciprocity during interactions (behavior) and the area of cognitive flexibility, manifesting itself in the difficulty of managing changes and transitions, abstracting information, presence of restrictive interests and/or sensory alterations (American Psychiatric Association 2014). Other key aspects, such as autonomy for activities of daily living or academic performance, are also conditioned by the executive dysfunction underlying autism, regardless of the person's intellectual capacity (Merino 2016).

2 Literature Review

For a long time, the existence of rigid patterns of behavior, thought or interest defining subjects diagnosed as ASD was considered a reflection of a deficit in creativity and imagination (Craig and Baron-Cohen 1999; Boucher 2007). However, nowadays, this interpretation is less frequent, and some authors consider that the existence of idiosyncratic patterns of thought common to ASDs is at the basis of growing creativity (Kasirer and Mashal 2014) and an essentially original thought (Asperger 1941). From an analysis focused on creativity linked to science, some studies consider the mind of the TEA as a creative mind with a special ability for the processes of systematization-categorization, great visual-spatial capacity, or extraordinary memory applied to the arts, engineering, and science-related disciplines (Baron-Cohen et al. 1997, 2007, 2011).

From a hereditary perspective, Lyons and Fitzgerald (2013) suggest that creativity in ASD individuals is highly heritable. Based on the way students with ASD process perceptive information, focused on details, filtering decontextualized information, and considering aspects that for others go unnoticed, this author states that students with autism spectrum disorder have an advantageous cognitive profile for creativity. In addition, Fitzgerald emphasizes ASD individuals' ability to focus attention for long periods without distractions and on the large capacity of reproductive and exact memory (Lyons and Fitzgerald 2013). Similarly, the divergent thought, the low inhibition, and the perceptive field independence of people diagnosed with ASD (Noens and van Berckelaer-Onnes 2005; Horlin et al. 2016) are considered to be aspects underlying the creativity construct. Finally, it has been considered that the specific skills of ASDs, such as a special memory to store knowledge or language (Happé and Frith 2010), or hypersensitivity to stimuli such as color or music (Bhatara et al. 2013), represent the basis of original thought and therefore allow these individuals to generate a novel vision concerning the norm.

Although the characteristics of ASD associated with creativity have been widely studied in the literature, there is a need for studies that address the development of creativity in individuals diagnosed with ASD. This lack of studies becomes especially visible when trying to find references to programs that specifically work and evaluate this cognitive competence with double exceptionality students. This scarcity of research and programs makes it especially important to develop experiences that help to understand and deepen this aspect. Understanding how to stimulate creativity in individuals with ASD and double exceptionality would support a global understanding of the factors involved in this construct (Lyons and Fitzgerald 2013; Quirici 2015).

3 Method

3.1 The Intervention Program

A pre-posttest quasi-experimental design with a control group was used in this pilot study. The intervention lasted 30 h and was implemented during 3-h weekly sessions over ten weeks. Each session was focused on learning concepts of sound physics through the project-based learning approach. Participants were confronted with a central problem that could be solved through the design of a project. Some sessions were enriched with the assistance of specialists from different branches, such as musicians, physicists, or sound engineers.

Existing literature stresses the importance of using teaching approaches that encourage students to explore the intersections and boundaries shared by different disciplines, thus fostering cooperation between participants and creativity, understood as the ability to ask questions and develop novel solutions (Marinova and McGrath 2004; Stauffacher et al. 2006; Allen et al. 2010). In this sense, project-based learning involves the development of a final product that justifies and makes learning meaningful. The selection of such a product involves developing negotiation, communication, decision-making, problem-solving, and structuring and planning skills among groups of students. The autonomy of the participants to regulate their work is the basis for the development of meaningful learning. In the *TECNOArtea* intervention program, students are placed at the highest level of autonomy; thus, they have to decide on a project that links knowledge of physics and music and the teacher develops supports to generate strategies for problem analysis or decision-making but does not direct students' opinions or resolution options. The teacher only represents a guide or a scaffolding figure for their learning process and provides them with tools to search for solutions. Thus, each session encourages the development of questions and interest by exposing students to unanswered stimuli that improve their interest and desire in learning about fundamental concepts of the physics of sound. In this sense, learning is eminently practical and tangible, and considers the special characteristics of information processing that have students with autism.

Participants addressed topics such as: What is the sound? How can we visualize the sound? How can we modify the frequency of the sound? What practical application does the physics of sound have in everyday problems? What artistic applications does the knowledge of the principles of sound have? Using problem-based approach, focusing especially on generative questions and inducing children to create their own investigations, participants created different projects. For example, based on the visit of a music group, called *Fetén Fetén*, that makes music with everyday objects, students created instruments with recycled material by reproducing some musical devices. With this project, they worked on concepts such as sound frequency, speed, as well as the relationship between them. In another of the projects, a microcontroller board (i.e., Arduino), an ultrasound transmitter/receiver, and a buzzer were used to work on concepts related to sound wave, velocity of propagation, and frequency. Table 1 provides a description of each session. An in-depth description of the intervention, as well as photos and videos about the projects developed by the participants, can be consulted here: https://autismoburgos.wixsite.com/tecnoartea2018/.

3.2 *Sample*

Forty-two students (seven girls) diagnosed with double exceptionality (ASD and high IQ) constituted the sample of this study, with an average age of the participants of 11.21 (SD = 3.85), ranging from 8 to 17 years old. Half of the students were enrolled in elementary education ($n = 24$; 57.1%) at the start of the intervention. Students were drawn from the *Asociación Autismo Burgos* institution situated in the city of Burgos (Spain). Students participated in the program voluntarily and most of them knew each other at the start of the intervention program. Fourteen students (two girls) participated in the *TECNOArtea* program and thus represent the treatment group. The remaining twenty-eight students (five girls) were assigned to a wait-list control group. Data of one student was removed from the treatment group for being a univariate outlier.

3.3 *Instrument for Data Collection*

Data was collected using the CREA-C test (Corbalán Berná et al. 2003), which is a valid and reliable cognitive measure of creativity that examines the ability of individuals to elaborate questions related to a graphic material as a procedure for the measurement of creativity. Students were shown the picture C of the test and were instructed to write for four minutes as many different questions that can be asked about what is depicted in the picture. The test, consistent with Sternberg's conceptualization of creativity adopted in this study, assumes that more questions reflect higher levels of creativity.

Table 1 Description of the intervention program

Session	Aim	Scientific content	Activities performed
1	Exploration of different musical instruments and their characteristics. Visit of the music group *Fetén Fetén*	Sound and vibration. The sense of hearing. The eardrum and the brain. Relationship between science and arts	Experimentation with strings-based instruments and resonance boxes. Problem formulation: can we see the sound produced by the instruments?
2	Sound pressure analysis. Using "entertainment software" to create music	Parameters of an acoustic wave: sound wave amplitude and pressure. The transmission of sound	Acoustic experiments. Taking measurements with mobile devices. Musical composition workshop
3	Visualization of sound. Design of scientific projects	Sound and kinetic energy. Introduction to "Chladni plates." The engineering design process	Design of musical instruments using common materials. Brainstorming of scientific projects
4	Understanding sound: high/low notes and deep/sharp sounds. Development of the scientific project	Parameters of an acoustic wave: amplitude, period, and frequency	Use of didactic simulators. Problem-based work
5	How musical instruments produce sound. Development of the scientific project	Environment-related conditions. Acoustic wavelength and speed of sound	Representation of waves with "Geogebra." Use of didactic simulators. Guided experimentation with everyday objects
6	Development of the scientific project	Sound waves	Collaborative and problem-based work
7	Development of the scientific project	Sound waves	Collaborative and problem-based work
8	Development of the scientific project: elaboration of a scientific poster	Content and structure of a scientific poster	Collaborative and problem-based work
9	Development of the scientific project: elaboration of a scientific poster and oral presentation practice	Oral presentation at science-related conferences. Public exposure techniques	Collaborative and problem-based work
10	Oral presentation at a science fair	–	Presentation of the project developed

3.4 Procedure for Data Analysis

Given that treatment and control groups differed greatly in their pretest scores, data were first analyzed using a one-way between-group analysis of covariance (ANCOVA) to determine whether there is a significant difference in creativity levels for participants in the treatment group and the control group while controlling for their pretest scores and their age. The independent variable was pedagogical condition (treatment vs. control group). The dependent variable was the posttest scores on the CREA-C instrument, measuring creativity development, administered following completion of the program in both the treatment and control group. Pretest scores on the CREA-C instrument, administered at the start of the program, were used as a covariate to control for individual differences. Besides, data were further analyzed using a two-way between-group ANCOVA by adding a second independent variable, mainly age (Group 1: elementary graders aged 8–11; Group 2: secondary graders aged 12–17).

Preliminary checks were conducted to examine assumptions for ANCOVA. Levene's test indicated equal variances ($F = 2.788, p = 0.103$), thus the assumption of homogeneity of variance was not violated. Likewise, the distribution of pretest and posttest scores for each pedagogical condition revealed linearity; thus, the assumption of linear relationship was also not violated. Finally, there was no interaction between the independent variable (treatment vs. control group) and the covariate (pretest scores) ($p > 0.05$); thus, the assumption of homogeneity of regression slopes was not violated, and therefore, ANCOVA results can be interpreted with confidence. Although the sample size is not very large, it is above the minimum of n quota $\geq$ 30, and assumptions test was met; hence, we opted for parametric instead of non-parametric statistics to be able to control for pretest scores using a robust test, such as ANCOVA, instead of multiple non-parametric testing, which would have increases likelihood of Type I error (false positive).

4 Results

After adjusting for pre-intervention scores, a one-way ANCOVA revealed that there was a significant difference between the treatment and the control group, $F(1, 39) = 37.06, p < 0.001, \eta_p{}^2 = 0.428$, indicating significant treatment effects with strong effect size (Fig. 1). The treatment group had significantly higher creativity scores for the adjusted posttest mean ($M_{\text{adjusted}} = 14.03$; SE = 0.88) relative to the control group ($M_{\text{adjusted}} = 8.23$; SE = 0.58), suggesting that the intervention program had significantly improved participants creativity. The independent sample t-test on the covariate (i.e., pretest scores) failed to show group difference, $t(40) = 1.670$, $p = 0.103$, indicating that the treatment and control group did not differ in their baseline scores collected at the start of the intervention program.

The two-way ANCOVA revealed that there was a significant interaction effect between the independent variables, $F(1, 37) = 7.775, p = 0.008, \eta_p{}^2 = 0.174$, indicating that participants enrolled in primary and secondary grades responded differently to the intervention (Fig. 2). Elementary graders from the treatment group had significantly higher creativity scores for the adjusted posttest mean ($M_{\text{adjusted}} = 16.67$; SE = 1.16) in comparison to the secondary graders from the treatment group (M_{adjusted}

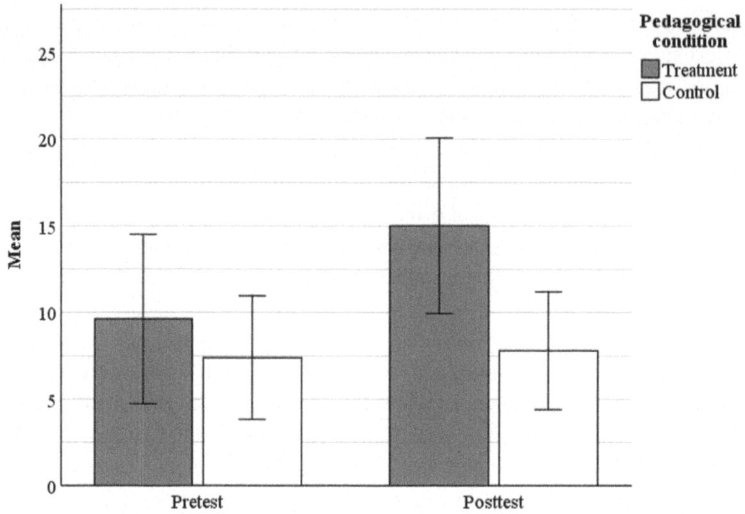

Fig. 1 Unadjusted means for the control and treatment groups

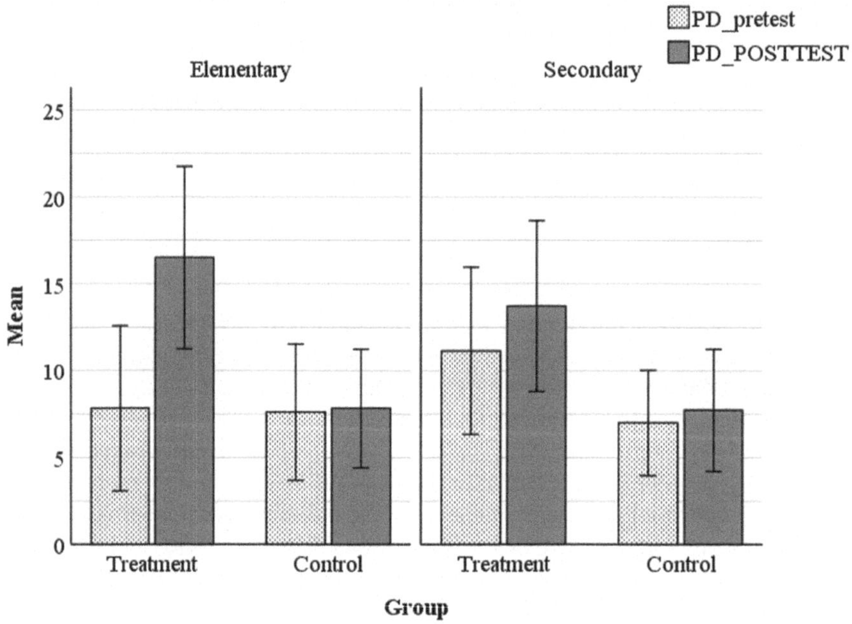

Fig. 2 Unadjusted means for the control and treatment group, according to school grade level

= 11.522; SE = 1.127), suggesting that the intervention program was more effective in developing creativity for students aged 8–11. However, a Mann–Whitney *U* test failed to show any difference between elementary and secondary school students pretest ($U = 13.00, p = 0.248$) and posttest scores ($U = 18.00, p = 0.667$); therefore, it can be concluded that the proposed program was effective in improving participants creativity levels, without differential effect based on participants age variable.

5 Discussion

This study aimed to design, implement, and evaluate a program for the development of the creativity of students aged 8–17 that have been diagnosed with ASD. The program consisted of an out-of-school, informal initiative that lasted for 30-total hours and delivered science content related to the physics of sound through project-based learning approaches. Taken together, the findings are consistent with existing literature that postulates that creativity is fostered through active, student-centered methodologies (Sousa and Pilecki 2013). The novelty of this study lies in the fact that we examined the effectiveness of these types of teaching approaches with an understudied population, namely students with double exceptionality (ASD and high IQ). The use of this methodology in students with autism and high capacity has been positive and has emphasized the importance of motivation, practical, functional, and significant teaching as essential to increase the interest and focus of thought (Schopler and Mesibov 2013).

Likewise, the use of sound physics as a motivating element has been especially useful and has allowed students to approach scientific concepts in a significant way, concretizing the theoretical contents approached in practical projects. Therefore, the results of the study indicate that problem-based learning is a useful strategy for developing the creativity of the students in this sample.

The program focused its hypothesis on the development of creative thinking as the basic skill to generate motivating questions and direction toward the development of innovative and congruent ideas. Future lines of research could focus on assessing the impact of an intervention as described in other aspects, such as collaboration between participants, their social competence, their concentration skills, or the improvement of affective-emotional aspects such as self-esteem or self-concept.

These findings reported in this study should be considered by taking into account the following strengths and limitations. On the one hand, in terms of strengths, the use of a valid and reliable instrument for data collection increases the confidence that can be placed in the results obtained. Likewise, the duration of the intervention has been long enough to allow observing a significant development of the creativity of the participants. Finally, it is also worth mentioning the strength of the methodological design, which used a control group with the same characteristics as the experimental group, thus, doting the study with internal validity that allows inferential conclusions to be drawn on the effectiveness of the proposed program. Concerning the limitations, it should be noted that fairly small sample size was used. Nonetheless, taking into account the special characteristics of the target population and the difficult access to this group, both the treatment and control group have been large enough to advance initial results about the effectiveness of extracurricular science programs, delivered using problem-based teaching strategies, in the development of double exceptionality students' creativity.

Acknowledgements We would like to thank the families of the students included in the sample, and to the *Asociación Autismo Burgos*, who have made this study possible. Likewise, this study was supported by the Spanish Foundation for Science and Technology (FECYT) under the FCT-17-12439 research grant.

References

Allen B, Crosky A, Yench E, Lutze-Mann L, Blennerhassett P, Lebard R et al (2010) A model for transformation: a trans-disciplinary approach to disseminating good practice in blended learning in science faculty. In: ASCILITE 2010—the Australasian society for computers in learning in tertiary education, pp 36–48

American Psychiatric Association (2014) Guía de consulta de los criterios diagnósticos del DSM-5: Spanish edition of the desk reference to the diagnostic criteria from DSM-5. American Psychiatric Publishing, Spain

Asperger H (1941) Zur Erziehungstherapie in der Jugendfürsorge. Austria. Monatsschr Kinderheilkd 87:238–247

Baron-Cohen S, Wheelwright S, Stott C, Bolton P, Goodyer I (1997) Is there a link between engineering and autism? Autism 1:101–109

Baron-Cohen S, Wheelwright S, Burtenshaw A, Hobson E (2007) Mathematical talent is linked to autism. Hum Nat 18:125–131

Baron-Cohen S, Ashwin E, Ashwin C, Tavassoli T, Chakrabarti B (2011) The paradox of autism: why does disability sometimes give rise to talent. In: Kapur N (ed) The paradoxical brain. Cambridge University Press, Cambridge, pp 274–288

Bhatara A, Quintin EM, Fombonne E, Levitin DJ (2013) Early sensitivity to sound and musical preferences and enjoyment in adolescents with autism spectrum disorders. Psychomusicology 23:100

Boucher J (2007) Memory and generativity in very high functioning autism: a firsthand account, and an interpretation. Autism 11:255–264

Chan ZC (2013) Exploring creativity and critical thinking in traditional and innovative problem based learning groups. J Clin Nurs 15–16:2298307

Corbalán Berná FJ, Martínez Zaragoza FA, Donolo DS, Alonso Monreal C, Tejerina Arreal M, Limiñana-Gras RM (2003) CREA. Inteligencia Creativa. Una medida cognitiva de la creatividad. TEA, Madrid

Craig J, Baron-Cohen S (1999) Creativity and imagination in autism and Asperger syndrome. J Autism Dev Disord 29:319–326

European Commission (2015) Science education for responsible citizenship. European Commission, Brussels

Fekula MJ (2011) Managerial creativity, critical thinking, and emotional intelligence: convergence in course design. Educ Innov J 3:92–102

Happé F, Frith U (2010) Autism and talent. Oxford University Press, Oxford

Horlin C, Black M, Falkmer M, Falkmer T (2016) Proficiency of individuals with autism spectrum disorder at disembedding figures: a systematic review. Dev Neurorehabil 19:54–63

Kasirer A, Mashal N (2014) Verbal creativity in autism: comprehension and generation of metaphoric language in high-functioning autism spectrum disorder and typical development. Front Hum Neurosci 8:615

Lyons V, Fitzgerald M (2013) Critical evaluation of the concept of autistic creativity. In: Fitzgerald M (ed) Recent advances in autism spectrum disorders. IntechOpen, Oxford, pp 771–791

Marinova D, McGrath N (2004) A transdisciplinary approach to teaching and learning sustainability: a pedagogy for life. In: Seeking educational excellence, proceedings of the 13th annual teaching and learning forum. Murdoch University, Murdoch

Marrapodi J (2003) Critical thinking and creativity: an overview and comparison of the theories. Critical thinking and adult. Providence, RI. Unpublished ED7590

Merino M (2016) Diseño y validación de un programa de habilidades para la vida en adolescentes con trastornos del espectro del autismo. Universidad de Burgos, Burgos

NGSS Lead States (2013) Next generation science standards: for states, by states. Academies Press, Washington, DC

Noens IL, van Berckelaer-Onnes IA (2005) Captured by details: sense-making, language and communication in autism. J Commun Disord 38:123–141

Piawa CY (2010) Building a test to assess creative and critical thinking simultaneously. Procedia Soc Behav Sci 2:551–559

Quirici M (2015) Geniuses without imagination: discourses of autism, ability, and achievement. JLCDS 9:71–88

Runco MA (2014) Creativity. Academic Press, Boston

Schmidt HG (1983) Problem-based learning: rationale and description. Med Educ 17:11–16

Schopler E, Mesibov GB (2013) Learning and cognition in autism. Springer Science & Business Media, New York

Snyder LG, Snyder MJ (2008) Teaching critical thinking and problem solving skills. J R Bus Educ 50:90

Sousa DA, Pilecki T (2013) From STEM to STEAM: using brain-compatible strategies to integrate the arts. Corwin Press, California

Stauffacher M, Walter AI, Lang DJ, Wiek A, Scholz RW (2006) Learning to research environmental problems from a functional socio-cultural constructivism perspective: the transdisciplinary case study approach. Int J Sustain High Educ 7:252–275

Sternberg RJ, Lubart TI (1995) Defying the crowd: cultivating creativity in a culture of conformity. FPF, New York

Takeuchi H, Taki Y, Sekiguchi A, Nouchi R, Kotozaki Y, Nakagawa S et al (2014) Creativity measured by divergent thinking is associated with two axes of autistic characteristics. Front Psychol 5:921

3D-Printed Plasma Cathode Electron Source for Educational Purposes

Fabian Bernstein, Sascha Schmeling, Thomas Wilhelm, and Julia Woithe

Abstract In secondary education, cathode-ray tubes (CRTs) are often the first choice when it comes to investigating the behavior of electrically charged particles in electric and magnetic fields. While CRTs offer some advantages, mainly from a practical point of view, they are on the whole ill-suited for an inquiry-based approach since they provide very limited room for modification or hands-on experimentation. Therefore, a 3D-printable plasma electron gun has been developed, which is at the same time modular, inexpensive, and easily accessible. The main objective is to provide teachers and students with an easy-to-operate electron beam source that allows conducting experiments on beam generation, beam focusing, and beam deflection as a hands-on activity in a classroom setting. From a technical point of view, this can be achieved by substituting the hot cathode as standard electron source by a plasma cathode electron gun, which can be operated at fore-vacuum pressure and in reactive gases. We provide a proof of concept that a low-cost 3D-printed plasma electron source is feasible and that beam generation can be accomplished.

Keywords 3D printing · Electron source · Hands-on experiment

F. Bernstein (✉) · S. Schmeling · J. Woithe
CERN, 23, 1211 Geneva, Switzerland
e-mail: fabian.bernstein@cern.ch

S. Schmeling
e-mail: sascha.schmeling@cern.ch

J. Woithe
e-mail: julia.woithe@cern.ch

F. Bernstein · T. Wilhelm
Department for Physics Education Research, Goethe University Frankfurt, Max-von-Laue-Str. 1, 60438 Frankfurt, Germany
e-mail: wilhelm@physik.uni-frankfurt.de

B. Jarosievitz and C. Sükösd (eds.), *Teaching-Learning Contemporary Physics*, Challenges in Physics Education,
https://doi.org/10.1007/978-3-030-78720-2_18

1 Motivation

The deflection of electrically charged particles in magnetic and electric fields is a core concept in physics and central to many physics curricula worldwide (Institute of Physics 2021).[1] Experimentally, cathode-ray tubes (CRTs) have long been the standard for studying this subject in a classroom setting. This might be due to the fact that cathode-ray tubes offer numerous advantages, ranging from relative robustness and quick and easy setup to reliable and convincing reproduction of the phenomena of interest. In addition, CRTs allow for either qualitative or quantitative experimentation. Upon closer examination, however, doubts arise as to whether these primarily practical advantages might not be bought by severe disadvantages from an educational point of view: for instance, cathode-ray tubes do not form part of students' daily lives, thus raising the long-disputed issue of relevance and interconnection of out-of-school experience with school learning (Williams et al. 2003). The discussion dates back as far as John Dewey, who identified the "gap existing between the everyday experiences of the child and the isolated material supplied in such large measure in the School" (Stuckey et al. 2013) as a major obstacle to student learning. It could thus be argued that the static and artificial nature of cathode-ray tubes stresses the detachment of what is taking place in the physics classroom to what matters outside of it and, therefore, rather inhibits than facilitates meaningful learning.

From a methodical point of view, cost constraints often inhibit the use as a hands-on experiment carried out by students, thus resulting in teacher demonstrations leaving the students as mere bystanders. Ideally, however, didactical considerations would guide the lesson planning rather than inversely financial constraints dominating the educational setting. Even more severely, the very construction of cathode-ray tubes prohibits any closer examination or even disassembly of the components without rendering the device unusable. As a result, CRTs are more likely to be perceived as "black boxes," prohibiting open inquiry and hindering any substantial understanding of the underlying physical principles.

To overcome these educational drawbacks of CRTs, we developed a novel 3D-printed plasma cathode electron source with the objective of making it as affordable, modular, and easily accessible as possible and thus allow for an in-depth and hands-on investigation of charged particle beams. This has largely been achieved by substituting the hot cathode by a plasma cathode, which relies on a gas discharge rather than thermionic emission as electron source, and by 3D-printing all the main system components to resolve the conflict between economic, technical, and educational boundary conditions.

[1] Some of the ideas in this article have been discussed in Bernstein et al. (2019).

2 Fundamental Considerations

Technically, cathode-ray tubes rely on thermionic electron sources, which are inexpensive, fully comprehended, and easy to control. On the downside, such electron sources can only be operated at a pressure of at most 10^{-5} mbar or in non-reactive gases (Oks and Schanin 1999). This is why cathode-ray tubes for educational purposes typically contain rare gases at a pressure of approx. 10^{-1} mbar, which simultaneously serves the purpose of focusing the electron beam and making it visible. From this follows, however, that the tube cannot be opened, disassembled, or modified, which in turn prohibits the construction of a modular system.

For this very reason, the design of the low-cost electron source has been based on a plasma electron emitter instead, which operates by igniting a gas discharge and extracting electrons therefrom. Among other things, plasma electron sources come with the advantage of tolerating comparatively high pressure regimes of ca. 10^{-1} mbar (Burdovitsin et al. 1997). As a result, the individual components of the electron source can be made accessible for the students, while at the same time, the need for sophisticated vacuum technology, which is usually not available in schools, is removed.

3 Design and Optimization

3.1 Operating Principle of the Plasma Electron Source

The schematics of the plasma electron source are shown in Fig. 1. It consists of three electrodes in a recipient containing air at a pressure of approx. 10^{-1} mbar. If a potential is applied between anode and cathode, electrons will be accelerated towards the anode. Through inelastic collision with the gas molecules, new electron–ion pairs are generated, and hence new charge carriers become available to the process. Also, ions drifting to and finally striking the cathode can dislodge additional electrons via secondary emission. Generally speaking, the discharge will be self-sustaining if, on average, each free charge carrier creates at least one new charge carrier as replacement. In this case, the characteristic glow can be observed, which stems from excitation collisions of electrons with gas molecules (Demtröder 2009; Stroth 2011).

The underlying principle of electron beam generation from a glow discharge is the extraction of electrons from the plasma region using an extractor electrode, which is on positive potential relative to the anode. However, the main challenge with this design lies in the fact that the required accelerating field between anode and extractor can cause breakdown within the acceleration gap, in which case no beam extraction can be achieved. While it is possible to counteract the breakdown by establishing a pressure difference between the plasma generation region and the extraction region through differential pumping, this would, at the same time, nullify the advantages of design simplicity (Oks 2006).

Therefore, other measures have to be taken, primarily by making use of the peculiarities of the Paschen curve. While the breakdown voltage depends in complex ways on multiple parameters, such as gas type, electrode spacing, electrode shape, geometry of the recipient and cathode material, a simplified description of the breakdown voltage as a function of pressure and electrode spacing is given by the Paschen curves (Fig. 2) (Demtröder 2009). It shows that in a given gas, the breakdown voltage solely depends on the product of pressure p and electrode spacing d, the so-called Paschen parameter $p{\cdot}d$. In general, the breakdown voltage decreases, if either the pressure is lowered or the spacing between the electrodes is reduced. Surprisingly, however, Paschen curves show a minimum, i.e., from a certain distance on, a further reduction of the electrode spacing will lead to an increase of the breakdown voltage. The prime strategy for increasing the breakdown voltage and thus the possible accelerating potential is, therefore, to choose a sufficiently small gap between anode and extraction electrode. In practice, the accelerating voltage, electrode spacing, and beam emittance are intertwined, which is why further analysis and simulation of the system dynamics are required.

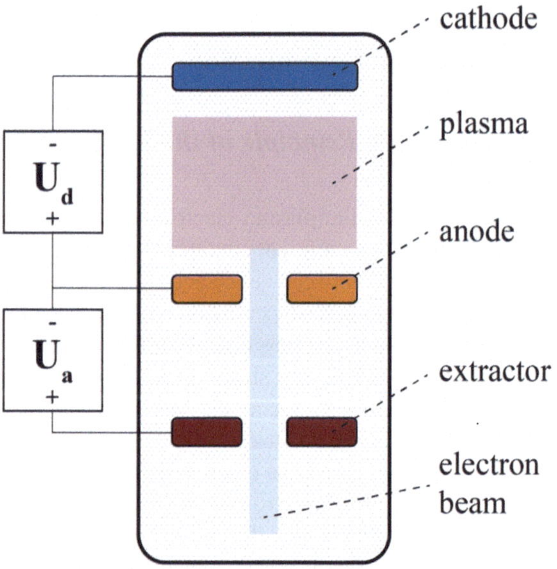

Fig. 1 Schematic of the plasma electron source

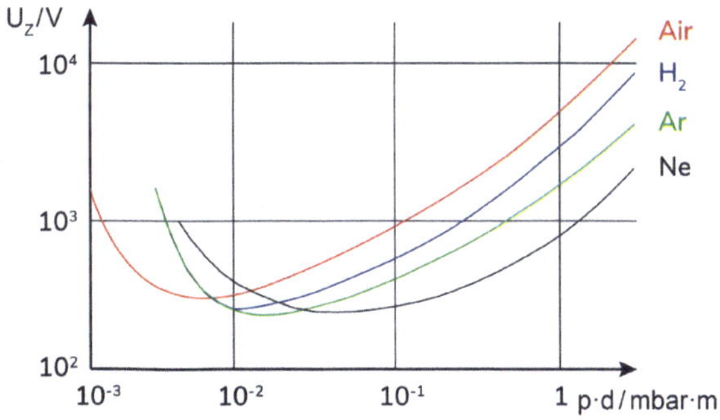

Fig. 2 Paschen curves of various gases (based on Demtröder 2009)

3.2 Design of the Plasma Electron Source

The overall system design is depicted in Fig. 3.

To reduce costs, 3D-printed parts (ABS on standard FDM printer) and standard hardware store components (screws, spacers, O-rings) have been used exclusively. An aluminum cylinder features as the cathode, whereas anode and extractor are made from steel spacers. Electric connection is achieved by screws that simultaneously seal the cap. The assembly is straightforward and requires no tools. As recipient, a standard plexiglass tube has been used, the connection for the vacuum hose also being 3D-printed. Special attention has to be paid to the belt tension of the FDM printer—poor calibration will lead to elliptical rather than round caps, which consequently complicates the sealing of the tube. By using only an inexpensive rotary vane pump, a pressure of ca. 10^{-1} mbar can be obtained. The ignition of a gas discharge as a prerequisite for a plasma electron gun is effortlessly possible, as can be seen in Fig. 4.

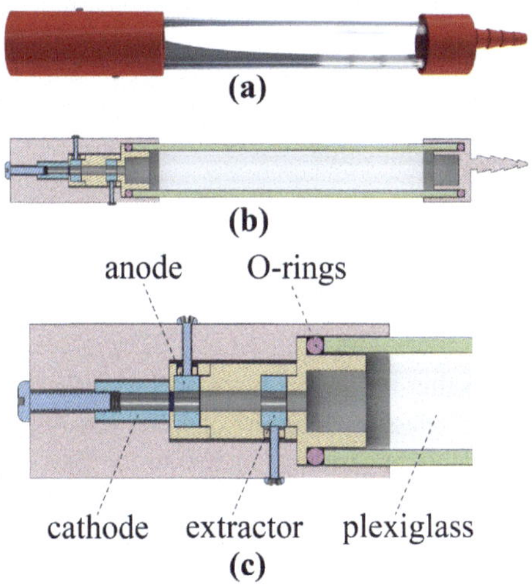

Fig. 3 Design of the 3D-printed plasma electron source: **a** overview with hose connector, **b** section analysis, **c** section analysis (detail)

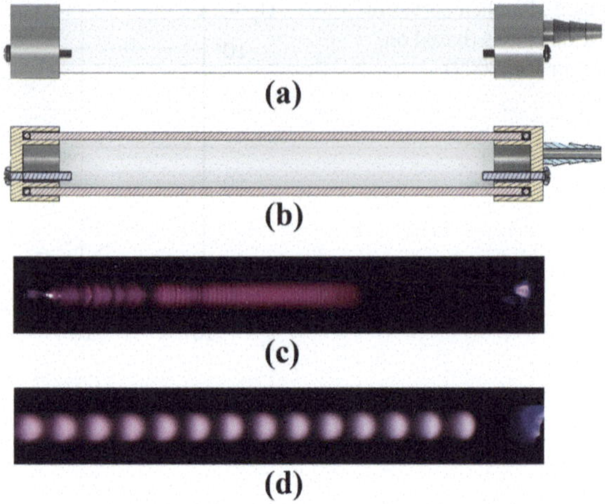

Fig. 4 Design of a 3D-printed discharge tube consisting of a plexiglass tube and two 3D-printed end caps: **a** overview, **b** section analysis, **c**, **d** characteristic glow discharge in air at a pressure of ~5 mbar (**c**) and ~10^{-1} mbar (**d**)

3.3 Simulation and Optimization

The extraction of an electron beam from the plasma, which is suitable for classroom experimentation, is technically more demanding. Due to the complex dynamics of the plasma and the interdependence of the various system parameters, simple qualitative considerations are not sufficient. Therefore, simulations of the electric potential and electric field have been performed in "Comsol" for multiple geometries and varying potential differences to reduce the number of setups that had to be evaluated experimentally.

Figure 5, for example, shows that the electric potential is not symmetrical due to the geometry of the electron source and the indispensable electrical connections, which has a direct impact on the beam propagation, as can be seen in Fig. 6.

A more thorough analysis also reveals that the divergence of the electron beam is directly tied to the spacing between cathode, anode, and extractor, while at the same time the spacing between anode and extractor accounts for the maximum acceleration voltage achievable.

4 Conclusions

Up to this point, we have accomplished a proof-of-principle that a low-cost 3D-printed plasma electron source is feasible and that beam generation can be accomplished. The residing air inside the plexiglass container renders the electron beam visible, as can be seen in Fig. 7. Spectral analysis further reveals that the primary peaks which account for the bluish tone are centered around 391 and 427 nm (Fig. 8) and stem from electron excitation of nitrogen (Ave and Bohacova 2007).

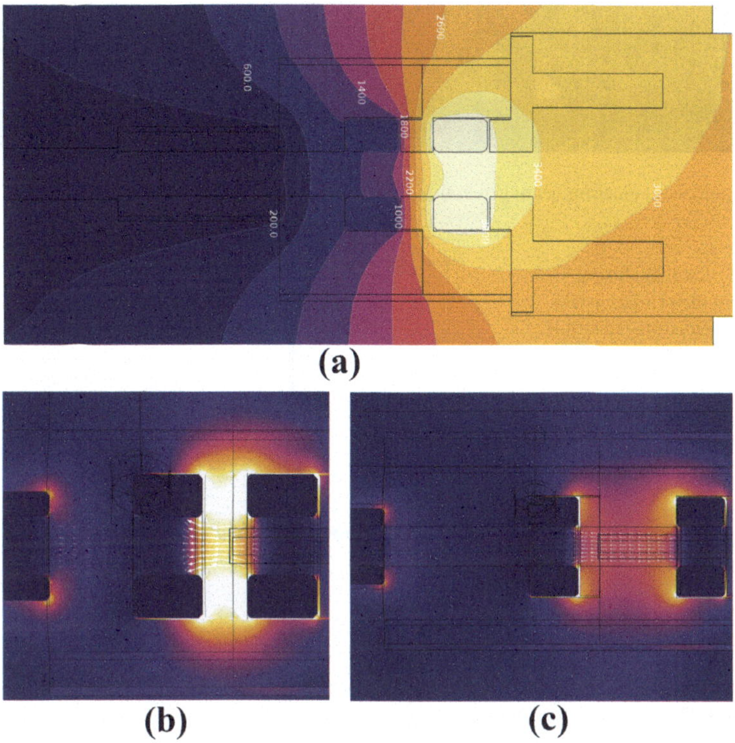

Fig. 5 Simulation of the electric potential (**a**) as well as the electric field (**b**, **c**) for different designs. Spacing cathode-anode: 6 mm (**b**), 15 mm (**c**) spacing anode extractor: 3 mm (**b**), 10 mm (**c**)

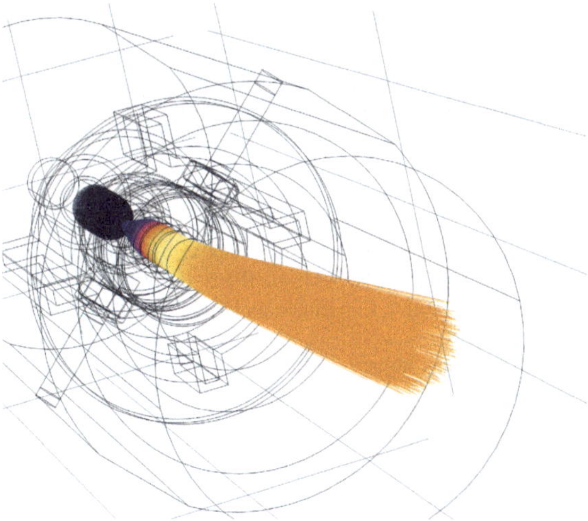

Fig. 6 Simulation of the electron beam propagation through the source (spacing cathode-anode 6 mm, anode-extractor 3 mm)

Fig. 7 Successful electron beam generation in the 3D-printed plasma cathode electron gun

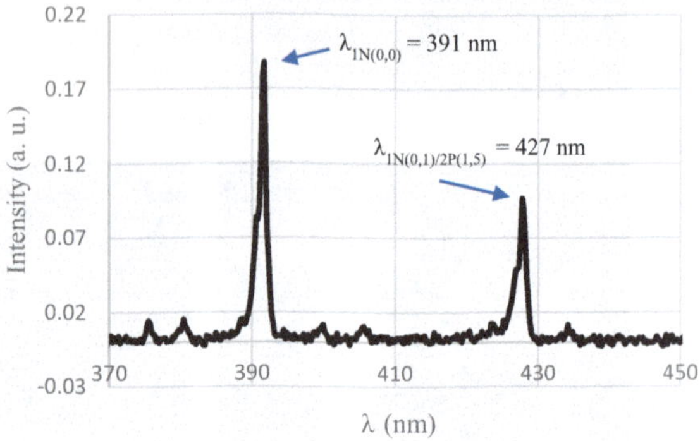

Fig. 8 Spectral analysis shows that the primary peaks stem from electron excitation of N_2

However, further advances have to be made to improve the overall beam quality and reliability before the plasma electron gun can be a valid alternative to CRTs, truly allowing advanced electron beam manipulation in a classroom setting.

References

Ave M, Bohacova M et al (2007) Measurement of the pressure dependence of air fluorescence emission induced by electrons. Astropart Phys 28:41–57

Bernstein F, Schmeling S, Wilhelm T (2019) Eine 3D-gedruckte Plasmaelektronenquelle für den Physikunterricht. In: Nordmeier V, Grötzebauch H (eds) PhyDid B, Didaktik der Physik, Beiträge zur DPG-Frühjahrstagung Aachen 2019, pp 249–252

Burdovitsin V, Oks EM, Serov A (1997) Hollow cathode electron gun for beam generation in forevacuum gas pressure. In: Markovits M, Shiloh J (eds) 12th international conference on high-power particle beams BEAMS98. IEEE, pp 412–415

Demtröder W (2009) Experimentalphysik, Vol. 2, Elektrizität und Optik, 5th edn. Springer, Heidelberg

Institute of Physics. https://spark.iop.org/collections/electron-beams-cathode-rays. Accessed 13 Jan 2021

Oks EM (2006) Plasma cathode electron sources. Physics, technology, applications. Wiley-VCH, Wiesbaden

Oks EM, Schanin PM (1999) Development of plasma cathode electron guns. Phys Plasmas 6(5):1649–1654

Stroth U (2011) Plasmaphysik. Phänomene, Grundlagen, Anwendungen. Vieweg+Teubner, Wiesbaden

Stuckey M, Hofstein A, Mamlok-Naaman R, Eilks I (2013) The meaning of 'relevance' in science education and its implications for the science curriculum. Stud Sci Educ 49(1):1–34
Williams C, Stanisstreet M, Spall K, Boyes E, Dickson D (2003) Why aren't secondary students interested in physics? Phys Educ 38(4):324–329

The Gecko® Approach to Friction: A Novel Teaching Learning Sequence

Cinzia Scorzoni, Guido Goldoni, and Valentina De Renzi

Abstract We here propose a novel teaching–learning sequence on tribology, based on the experimental investigation of Gecko® Nanoplast®, a bio-inspired micro-structured synthetic material with peculiar tribological properties. Measurements of static and dynamic friction show that Gecko® does not obey to Leonardo-Amontons' law, thus providing a way to introduce the key concept of real—as opposed to nominal—contact area. As in 'real' material-science research, investigation of the Gecko® macroscopic behavior is combined with its structural determination, based on both optical microscopy observations and on diffraction measurements. The sequence has been tested on a group of honors high-school students with encouraging results.

Keywords Friction · Nanoscience · Structural properties

1 Introduction

The term tribology refers to the area of research investigating those phenomena which take place at the interface between two bodies in intimate contact and in relative motion, i.e., friction, adhesion, wear, and lubrication. In the last decades, novel experimental and theoretical methods have disclosed new insight into the origin of tribological phenomena at the micro- and nano-scale (Busham et al. 1995; Carpick and Salmeron 1997). This complex and partially unexplored area of research has a strong interdisciplinary character, involving physics, chemistry and engineering, and has a huge technological impact (Jost 1966). For these reasons, this topic is particularly suitable to foster pupils' interest in unravelling the phenomena occurring at the micro- and nanoscale.

While in standard physics course little attention is usually devoted to friction, with seldom any attempt to explain its microscopic origin, in recent years a few

C. Scorzoni · G. Goldoni · V. De Renzi (✉)
Dipartimento di Scienze Fisiche, Informatiche e Matematiche, Università di Modena e Reggio Emilia and Istituto NANO CNR, Via Campi 213/A, 41125 Modena, Italy
e-mail: vderenzi@unimore.it

B. Jarosievitz and C. Sükösd (eds.), *Teaching-Learning Contemporary Physics*, Challenges in Physics Education,
https://doi.org/10.1007/978-3-030-78720-2_19

works have brought tribological phenomena to the attention of STEM teachers (Krim 2002). Indeed, their presentation in standard physics courses has been profoundly and critically revised, and new didactic approaches to this topic have been proposed (Corpuz and Rebello 2011a, b; Besson et al. 2007, 2010; Montalbano 2014).

Stemming from these works, we here propose a complementary and innovative approach based on the experimental investigation of the macroscopic and microscopic properties of Gecko®, a synthetic material which mimics the extraordinary adhesive properties of living Geckos.

The proposed teaching–learning sequence provides high-school students with an accurate, though simplified, description of the basic physics underlying friction. It has been proposed to and tested on a group of high school honors students, during a one-week stage at our department. Its effectiveness has been probed by comparing pre- and post-test results, showing very encouraging results, in particular as far as key issues as nanofriction, surface, and atomic interactions are concerned.

The paper is organized as follows: Sect. 2 quickly surveys the main concepts in tribology, as viewed in particular in the context of nanoscience and nanotechnology, and illustrates the properties of both the living Gecko and its artificial counterpart, i.e., Gecko®. In Sect. 3, we briefly review the literature most relevant for this work and present our didactic choices. The teaching–learning sequence is described in Sect. 4 and discussed in Sect. 5, along with its more significant results. In Sect. 6, some conclusions and perspectives are drawn.

2 Scientific Background

2.1 Tribology at the Micro and Nanoscale

Understanding the frictional behavior of solids is a very complex task, as it needs to consider a variety of different mechanism (as for instance adhesion, deformation, abrasion, the effect of absorbed layers), working at different scales, from the meso-down to the nano-scale. Despite this complexity, recent developments in both theoretical and experimental approaches have allowed to obtain a quite comprehensive picture of what occurs at the interface between two solids in intimate contact and relative motion. While we refer the readers to Busham et al. (1995), Carpick and Salmeron (1997), Krim (2002) for a comprehensive review of the subject, we here want to emphasis that a key concept to understand the microscopic origin of friction is that when two surfaces touch each other, the actual microscopic area of contact is much less than the apparent macroscopic area. As friction derives from the interactions between atoms and molecules of the actual contact areas, a key issue is to determine how this contact area should be defined. As molecular interactions occur at the nanoscale, only those atoms/molecules which are actually separated by few nanometers across the interface contribute to the actual contact. Moreover, the strength and the range of the interaction critically depends on the chemical nature of the interface (and on the possible presence of an intermediate layer, the so-called third body).

Geckos' Secret

As it is well known, Geckos are extraordinary climbers: they can run across walls and ceilings, carrying loads as heavy as ten times their weight; in doing so, their toes attach and detach from any surface within few milliseconds. Researchers have demonstrated that Geckoes can climb on whatever surfaces, no matter which material and roughness it presents (Autumn 2006). Moreover, they ruled out the presence of glue-like secretion, which may in principle explain this extraordinary effective adhesion. Indeed, as shown in Fig. 1, each of Gecko's toes is formed by thousands of micrometric *lamellae*, each of them formed by *setae*, which apex are furthermore structured down to the nanoscale. This hierarchical structure guarantees a huge contact area between the toe and the surface.

Due to this, all the tiny Van der Waals forces acting between each portion of Gecko's toe and the wall surface sum up building an overall adhesive interaction, large enough to overcome its weight. Moreover, by changing the angle of contact between the toes and the surface, the real contact area can be varied, thus enabling the rapid attaching/detaching from the surface.

Inspired by Geckos, several materials with novel adhesive properties have been designed and manufactured in recent times. Among these, Gecko® Nanoplast® film is easily purchased and reasonably cheap, being therefore well-suited for classroom

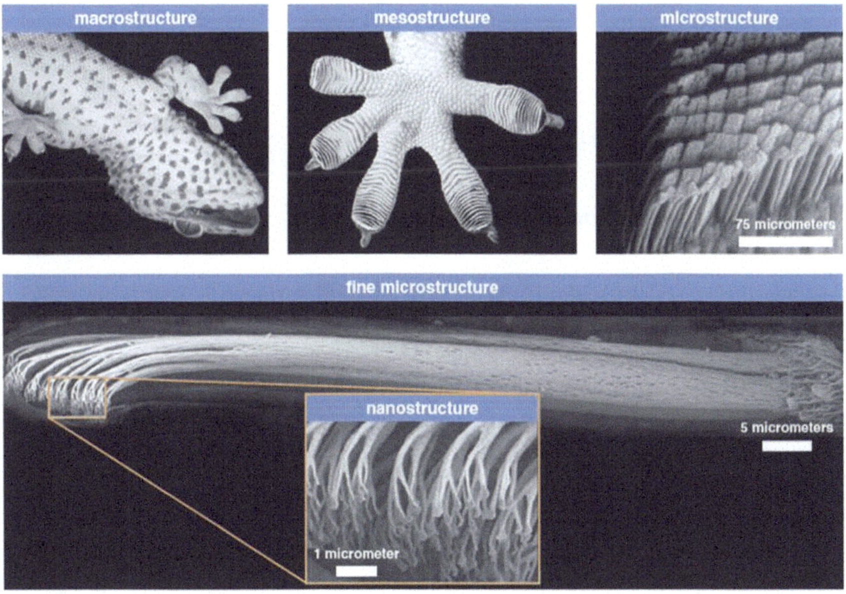

Fig. 1 Hierarchical structure of Gecko's feet, from the meso-scale, down to the nano-scale (taken from Wikipedia, under creative common permission)

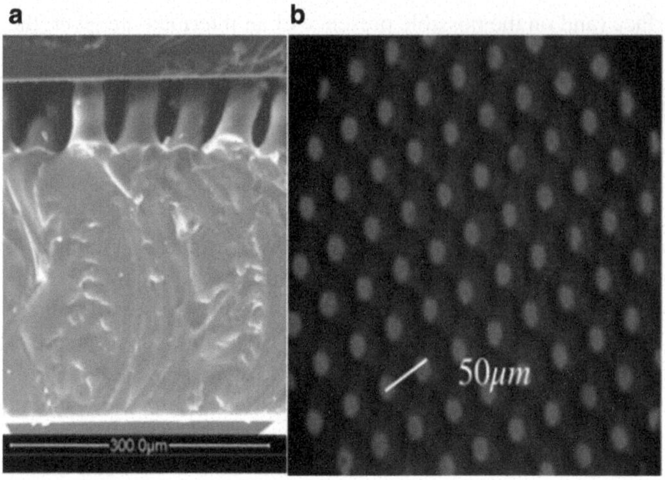

Fig. 2 **a** SEM lateral view of the Gecko®/silicon interface and **b** image of Gecko® micro-structured surface. The overall film thickness is 400 μm, and the hexagonal pattern is formed by 50 μm high pillars protruding from one side of the film, with lattice parameter of 50 μm

investigation. As shown in Fig. 2, one of its side is characterized by an ordered array of micropillars, cast out from a flexible thin film, while the other is flat. The micro-structuration and the flexibility of the material makes it suitable for adhesion on different kinds of surfaces, allowing for interesting application in both medical and aerospace (i.e. at low atmospheric pressure) fields.

3 Didactic Framework

3.1 *Friction in Schools*

In standard physics curricula, friction between solid surfaces is usually introduced as one of the existing different types of forces, together with elastic forces, weight, tension, and sometimes, viscous forces. Each of these forces are quantitatively defined by simple equations and are usually combined in a variety of manners in exercises designed to probe students' understanding of Newton's laws and conservation principles. This simplistic approach has been recently deeply criticized, as it hinders any true understanding of the actual phenomena occurring at the microscopic interface between sliding bodies (Corpuz and Rebello 2011a, b; Besson et al. 2007, 2010; Montalbano 2014). In particular, the importance of providing students with correct—though possibly simplified—structural models, has been highlighted, showing how it may facilitate the building of effective mental models of the underlying microscopic phenomena (Corpuz and Rebello 2011a, b; Besson et al. 2007,

2010). It is in particular relevant to: (a) clarify the distinction between nominal and real area of contact; (b) propose effective visual representation of meso- and micro-asperities in contact and explain their essential role in determining the tribological properties of interfaces; (c) introduce the atomic and molecular interaction model to explain the origin of friction at the microscopic level. To this end, the TL sequence proposed by Besson and co-workers (2007) devised a set of qualitative and semiquantitative experiments addressing: (i) vertical friction forces; (ii) static and kinetic friction; (iii) the difference between real and nominal contact area; (iv) friction phenomena from the point of view of energy. Drawings are also used to illustrate the importance of surface topography on a micrometric scale, and mechanisms producing friction are discussed.

3.2 *Didactic Choices*

Stemming from these works, we here propose a novel TLS suitable for the last years of ISCED 3 formation (16–18 years old students), which guides students through quantitative experiments, meant to disclose the link between macroscopic and microscopic properties of Gecko®. The proposed teaching sequence has been designed according to a few main didactic choices, as detailed in the following:

(i) From the theoretical point of view, our TLS takes inspiration from both the investigative science learning environment (ISLE) model developed by Etkina (2015) and the 5E (Engage, Explore, Explain, Elaborate, Evaluate) model by Bybee (2014). From the former, we took the idea of designing our sequence in a way that reproduces, as truly as possible, the modalities of actual scientific research. Students learn through quantitative, and experimental activities performed in groups, in a way that essentially resembles the so-called ISLE circle. In fact, at partial variance with this model, we here propose highly guided experiments, as the result of each experiment is meant to provide scaffolding for understanding the investigated phenomena and building correct mental models. As far as the 5E model is concerned, we particularly emphasize the initial activities which *Engage* students' interest, and the *Explain* activities, during which—exploiting modalities akin to those typical of scientific conferences—each group is asked to present and discuss with the peers the results of their own investigation.

(ii) As explained in (i), we intend to exploit pupils' interest in technology and energetic issues to *engage* their interest in friction and tribology. For this reason, rather than dealing with the importance of friction in everyday life, in the introductory lecture emphasis is put on the role of friction and, more in general, tribological issues in nowadays and future technology, trying also to give a flavor of the ongoing research topics.

(iii) Aiming to elucidate the concept of real contact area as opposed to the nominal (or geometric) contact area, we propose a quantitative investigation of the different behavior of "usual" materials (i.e., sandpaper) and Gecko®. The pieces of information thus obtained will be combined with those regarding Gecko® microstructure, providing a route towards the understanding of the microscopic mechanism at the basis of adhesion and friction.

(iv) Bridging the gap between the mental image of two macroscopic bodies in contact and that of the actual interface at the meso- and micro-scale is nothing but trivial. We here propose to exploit the optical microscope to observe the cross-section of different interfaces, thus helping to build a correct mental image of actual interfaces and of meso/micro-asperities in interaction.

4 The Gecko® Teaching–Learning Sequence

The description of Gecko's extraordinary capabilities—with no explanation of their origin—is initially used to nail pupils' interest and propose the experimental investigation of the animal artificial counterpart—i.e., Gecko® (GT in the following), in a series of qualitative and quantitative experiments. To further capture students' attention, biomimetics is also shortly introduced as a fascinating and smart way to envisage new technology. The proposed sequence starts with qualitative investigations on GT peculiar properties, its ultimate goal being the quantitative determination of its (i) macroscopic mechanical properties and (ii) microscopic structure. No preliminary knowledge of its properties and structure is provided.

4.1 Preliminary Observations

Initially, the GT adhesive properties are qualitatively tested on several surfaces, with different shape, roughness, and chemical composition. In particular, the difference between GT shear vs peeling behavior can be readily observed and compared with that of usual glue-based tapes. GT performances tend to degrade rapidly when it gets dirty, but—at variance with usual tapes—they can be easily re-established just washing it with tap water. More quantitatively, GT adhesion on vertical surfaces is probed using the simple set-up shown in Fig. 3a. GT is carefully positioned on a smooth vertical surface. The shear force it can sustain without detaching is probed by progressively adding weights to the clam. If carefully positioned, GT can sustain a shear force up to more than 10 N (it actually breaks rather than detaching). This experiment is designed to convey the key-idea that adhesion and friction are strictly related and provides an interesting example in which the normal force is not identified with weight. Moreover, it clearly illustrates the difference between peeling (the force required to peel GT from the vertical surface is negligible) and shear adhesion, allowing to introduce the concept of tuneable friction.

The striking difference between peeling and shear behavior, together with the results of simple manipulation and water-washing tests, easily brings students to rule

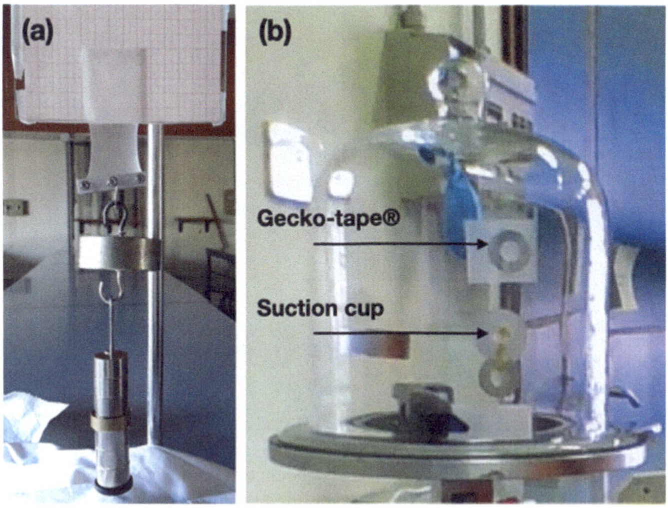

Fig. 3 **a** Experimental set-up to probe Gecko® vertical adhesion; **b** vacuum bell experiment showing the different mechanism of adhesion of Gecko® and a suction cup

out glue as a possible explanation for GT adhesion. At this stage, usually students propose the suction mechanism as an alternative explanation. In order to test this hypothesis, following the ISLE circle, a comparative experiment between suction-cup and GT adhesive behavior can be performed exploiting a vacuum bell, as shown in Fig. 3b. The experiment clearly shows that, while the suction cup (bearing a washer) detaches from the vertical frame once the pressure inside the bell is sufficiently low, the washer attached to GT remains in position. This clearly demonstrate the GT adhesion does not derive from a suction mechanism either. At this stage, the adhesive mechanism of GT remains therefore an open question, which engage students in further investigations.

4.2 *Static and Dynamic Friction*

The main goal of this set of measurements is to quantitative investigate the frictional behavior of GT and compare it with that of sandpaper (SP in the following), taken as an example of material with conventional properties. In particular, we are interested in probing the validity of Leonardo-Amontons' law, i.e., the independence of the frictional coefficients from the nominal area of contact, in the two cases. The standard set-up used in classrooms to measure stating friction consists of a variable-angle sliding plane and a test block. The test block is placed on top of the plane, and its angle gradually increased, till the block starts to slide downwards. The angle θ_S at which sliding starts is measured by means of a goniometer fixed at the vertex of the sliding pane. The friction coefficient is than readily determined as $\mu_S = \tan(\theta_S)$, by

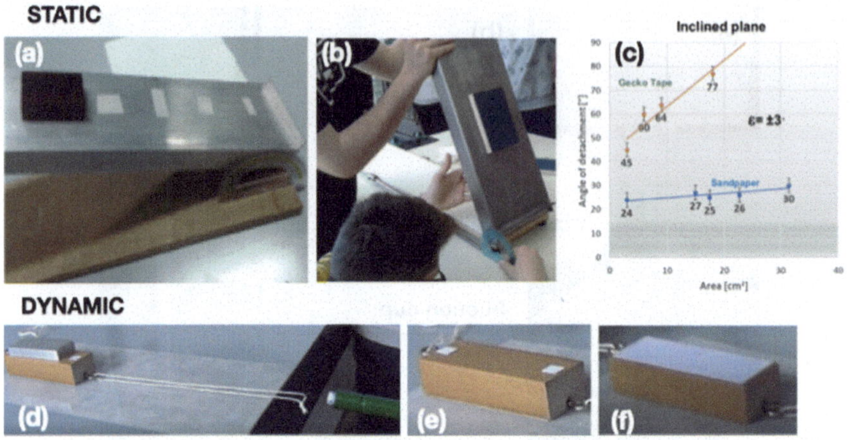

Fig. 4 **a, b** Experimental set-up for probing the dependence of static friction on the geometric area of contact; **c** The measured angle of detachment is reported as a function of the nominal contact area, for both Gecko® (orange points) and Sandpaper (light-blue points). The corresponding linear best fits are also shown. The uncertainty in angular determination is evaluated as $\pm 3°$. **d** experimental set-up for dynamic friction measurements: the different areas of contact of the two sides of the testing block are shown in **e** and **f**

applying force balance and the law of static friction. This set-up is typically used to test the validity of Leonardo-Amontons' law, varying the nominal contact area between the plane and the block by changing the block face that is in contact with the plane. As shown in Fig. 4a and b, to minimize the role of torque and peeling, we here use a different approach; we glue a few (at least three) pieces of SP with different areas to the sliding plane. The test-block is then carefully placed on top of each piece, so that the nominal contact area between the block and the coating piece coincides with that of the latter, and the sliding angle is measured. The same procedure is repeated using GT coating.

As easily derived from the measurements reported in Fig. 4c, in the case of SP the coefficient of static friction does not depend on the contact area, as expected. On the contrary in the case of GT, the coefficient of static friction clearly increases with the contact area, in apparent contrast with Amontons' law. This simple experiment highlights the peculiarities of GT frictional properties in a straightforward and striking way, fostering pupils' interest in unravelling their origin.

The frictional properties of GT can be further tested in dynamic conditions. To this aim, a standard set-up can be considered, as depicted in Fig. 4d, in which the dynamic frictional force of a wooden block dragged on a coated horizontal plane is measured by a dynamometer. Here, three different types of coatings are considered, i.e., sandpaper, GT exposing its flat surface and GT exposing its micro-structured one (see Fig. 2a). It is important here to point out that, while the two sides of GT can be easily distinguished by visual inspection—i.e., the flat one is glossy and the micro-structured one is opaque—students do not know at this stage what is the actual difference between the two. Indeed, while the different surface roughness may be somewhat perceived by running a fingernail on the tape, the micro-structuration definitely cannot be distinguished by naked eye. As shown in Fig. 4e, f, in order to vary the contact area between the test block and the surface coating, the two symmetric sides of the test block are covered with different areas of flat cardboard: in one case, the whole surface is covered, while in the other, only two small cardboard pieces (few mm^2) are glued to the block. As expected, in the case of SP, the measured dynamic friction does not depend on the block side, i.e., on the nominal contact area. The same is true also in the case of the glossy surface of GT. On the other hand, in the case of opaque GT surface, this dependence is quite remarkable. The completely different frictional behavior of the two sided of GT, which share the same chemical composition, is surprising and calls for further investigation.

4.3 Structural Properties

This part of the TLS aims at investigating the structural properties of GT, i.e., the details of its morphology. At this stage, students are therefore led to discover the link between the macroscopic tribological properties of GT and its structural properties. The latter can be investigated with two complementary approaches, described in the following sections.

Optical Microscope Investigation

Due to the dimension and distance of the micropillars (i.e., 50 μm), the microscopic structure of the GT surface can be easily revealed using a standard optical microscope. Optical microscopes are usually available in school labs and may also be USB connected to a computer to acquire high-resolution images. Alternatively, home-made microscopes which exploits the smart-phone optics may also be used.[1]

In panels (a) and (b) of Fig. 5, images of the surface and the cross-section of GT are reported. The GT surface can be observed both in reflection and in transmission, if GT is put on a transparent support. On the other hand, the best cross-section images can be obtained in transmission, by cutting a thin slice of GT (1 mm or less) and put it vertically on a transparent support. GT surface micrographs clearly show its hexagonal lattice structure, which periodicity can be easily estimated from the microscope magnification value. Inspection of the cross-section image (panel (b)) reveals that the ordered array is actually due to micropillars and may unravel the morphological difference between the two sides of GT, i.e., the opaque side micro-structuration and the glossy side flatness (not shown here).

[1] See for instance https://edu.workbencheducation.com/cwists/preview/6187-diy-cell-phone-microscopex.

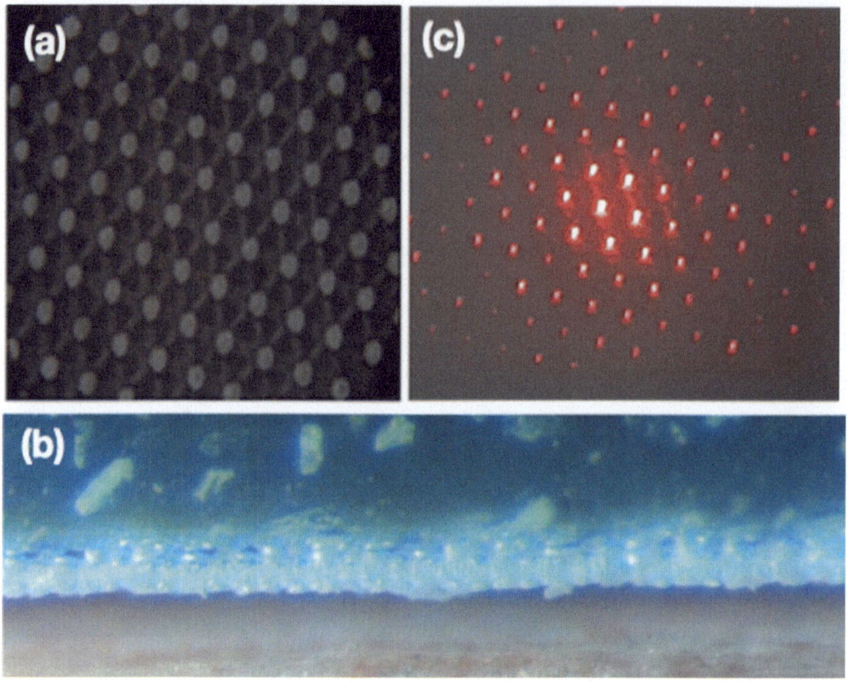

Fig. 5 Micrographs of **a** Gecko® surface and **b** the cross-section of the interface between Gecko® and a copper substrate. **c** Gecko® diffraction pattern, taken with a red laser-pointer

Diffraction Investigation

A complementary way to investigate the structural properties of the GT is by means of diffraction, as the ordered array of micropillars acts as a bi-dimensional hexagonal diffraction grating; furthermore, GT is almost transparent to visible light. As shown in Fig. 5c, by shining the light of a laser-pointer through a piece of GT film, a nice hexagonal diffraction pattern can be observed on a screen. By measuring the distance L between the GT and the screen and the distance d between neighboring spots, a quantitative estimate of the lattice parameter, i.e., the inter-pillar distance *a*, can be easily derived as

$$a = \frac{\lambda}{\sin\left(a\tan\left(\frac{d}{L}\right)\right)} \cong \frac{\lambda L}{d}, \tag{1}$$

where λ is the laser wavelength. The *a* value determined in the diffraction experiment is in good agreement with that obtained with the optical microscope.

4.4 Guided Discussion

At this stage of the sequence, the experimental results described in the previous sections are illustrated and collectively discussed in a plenary session. Aim of this session is to provide a direct link between the macroscopic properties of GT and its morphology, as compared with that of conventional materials, such as sandpaper. The discussion session can be performed with different modalities, depending on the time available and the number of students involved. In the case specifically considered here, of the one-week stage, students were divided into groups, each group performing one of the experiments in full details. Afterwards, students of each group explain to others—in a show-and-tell fashion resembling scientific fairs—their experimental set-up and the results of the experiment. This corresponds to the explain phase of the 5E model. Once each student has been getting to know the results of all experiments, the teaching sequence foresees a moment of collective discussion of all experimental findings and observations, during which students are brought to draw some conclusions on the microscopic origin of friction.

Putting together the information obtained by the macroscopic investigation with those obtained by microscopy and Gecko's toe images allows to clearly identify micro-structuration as the key element to explain GT properties. To help discussion, the teacher also provides students with the SEM images reported in Fig. 2, which give a more detailed picture of the GT structure. Moreover, images of Gecko's toe hierarchical structure can also be shown at this stage, highlighting the similarities between the living Gecko and GT. At this point, the difference between real and nominal contact area is naturally brought in, together with the concept that both friction and adhesion are related to the real contact area. The microscopic explanation of Amontons' law and the reason for its deviation in the case of micro-structured GT coating are thus also discussed.

The sequence further proceeds by introducing the basic concepts of intermolecular forces, which are usually at least in principle already known form the chemistry curriculum. This step highlights the interdisciplinary character of this topic, which is often quite a revelation for Italian students. An effective aid to delve into this issue has been provided by the PhET interactive simulation of the Colorado University Boulder.[2]

5 Results and Discussion

This sequence was tested on 31 honor fourth-year high-school students (17–18 years old), during a one-week stage. Students were divided in two homogeneous groups: the so called 'physicists' performed the whole one-week sequence, while the 'mathematicians' attended only a two-hour lecture, given by their 'physicist' peers at the

[2] Phet Colorado University Homepage https://phet.colorado.edu.

end of the week. For both groups, the performances were probed through questionnaires given as pre-test and post-tests. In analyzing the results, we divided questions in different groups, according to the topic addressed. As apparent in Fig. 6, for the 'physicists' group, performances increase significantly for all topics, with an average incremental increase in the percentage of correct answers of 21%. Remarkably, the sequence results particularly effective when key issues as nanofriction, surface, and atomic interactions are considered. Indeed, for these topics, the percentage of correct answers increases from an average of 37% in the pre-test to 74% in the final test.

Interestingly, a significant average increment of almost 14% was also obtained for the 'mathematicians,' suggesting the effectiveness of peer education sessions.

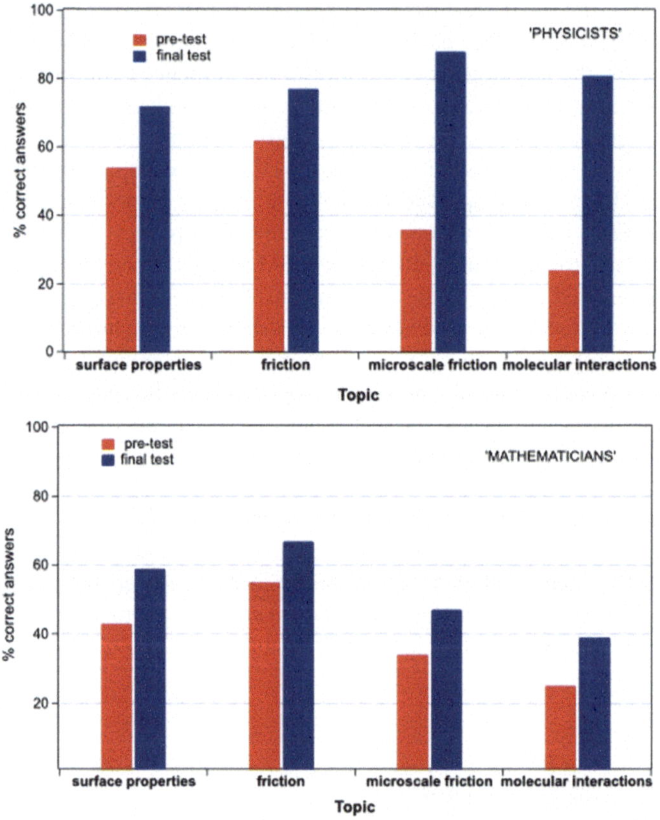

Fig. 6 Analysis of pre-test and post-test results for the two homogenous groups of honors high-school students

6 Conclusions

In conclusion, we here propose a novel TL experimental sequence, addressing friction through the investigation of the properties and behavior of Gecko®. The sequence has been proposed to a group of selected 4th grade pupils with homogeneous background, during a one-week university stage. The analysis of the results showed a significant increase in students' performances at the end of the proposed sequence, in particular as far as key issues as nanofriction and surface and atomic interactions are considered, suggesting its effectiveness in promoting students' understanding. During the testing, a few critical issues have been also highlighted, concerning in particular students' difficulties in realizing the actual dimensions of molecules, in creating a correct mental model of surfaces and solids at the mesoscale and in understanding the nature of molecular interactions. For these reasons at further completion of the proposed sequence, we are implementing a new set of experiments regarding the wetting properties of GT, whose effectiveness is currently been tested, with encouraging results.

References

Autumn K (2006) How Gecko toes stick. Am Sci 94:124

Besson U, Borghi L, De Ambrosis A, Mascheretti P (2007) How to teach friction: experiments and models. Am J Phys 75:1106

Besson U, Borghi L, De Ambrosis A, Mascheretti P (2010) A three-dimensional approach and open source structure for the design and experimentation of teaching-learning sequences: the case of friction. Int J Sci Educ 32:1289

Busham B, Israelachvili JN, Landman U (1995) Nanotribology: friction, wear and lubrication at the atomic scale. Nature 374:607

Bybee RW (2014) The BSCS SE instructional model: personal reflections and contemporary implications. Sci Child 51:8

Carpick RW, Salmeron M (1997) Scratching the surface: fundamental investigations of tribology with atomic force microscopy. Chem Rev 97:1163

Corpuz ED, Rebello NS (2011a) Investigating students' mental models and knowledge construction of microscopic friction. I. Implications for curriculum design and development. Phys Rev Phys Educ Res 7:020102

Corpuz ED, Rebello NS (2011b) Investigating students' mental models and knowledge construction of microscopic friction. II. Implications for curriculum design and development. Phys Rev Phys Educ Res 7:020103

Etkina E (2015) Millikan award lecture: students of physics—listeners, observers, or collaborative participants in physics scientific practices? Am J Phys 83:669

Jost P (1966) Lubrication (tribology)—a report on the present position and industry's needs. Department of Education and Science, H. M. Stationery Office, London

Krim J (2002) Resource letter: FMMLS-1: friction at macroscopic and microscopic length scales. Am J Phys 70:890

Montalbano V (2014) An inquiry-based laboratory on friction. In: ICPE-EPEC 2013 proceedings. MATFYZPRESS, Prague, pp 570–578

The manufacturer's authorised representative in the EU is Springer Nature Customer Service Centre GmbH, Europaplatz 3, 69115 Heidelberg, Germany. If you have any concerns regarding our products, please contact ProductSafety@springernature.com

Printed and bound by CPI Group (UK) Ltd, Croydon, CR0 4YY
30/09/2026
02225695-0007